A lexicon of aviation words and phrases,
with a special section on aircraft names

Canadian Cataloguing in Publication Data

Poteet, Lewis J.

　　Plane talk : push you, pull me : a lexicon of
aviation slang with a special section on aircraft names

　　ISBN 1-55207-002-6

1. Aeronautics - Slang - Dictionaires.　　I. Stone, Martin. II. Title.

TL509.P67 1997　629.13'003　　C97-941058-4

Interested readers may consult
our evolving catalogue on the Internet at
http://rdppub.com

Lewis J. Poteet & Martin J. Stone

(PUSH YOU, PULL ME)

A lexicon of aviation words and phrases,
with a special section on aircraft names

ROBERT DAVIES PUBLISHING
MONTREAL—TORONTO—PARIS

Copyright © 1997, Lewis J. Poteet & Martin J. Stone
ISBN 1-55207-002-6

Robert Davies Multimedia Publishing Inc.
330-4999 Saint-Catherine Street West,
Westmount, Quebec, Canada H3Z 1T3

This book may be ordered in Canada from
General Distribution Services:

☎ 1-800-387-0141 / ☎ 1-800-387-0172
📠 1-416-445-5967;

in the U.S.A. from General Distribution Services,
Suite 202, 85 River Rock Drive, Buffalo, N.Y. 14287
☎ 1-800-805-1082

or from the publisher, toll-free throughout North America:
☎ 1-800-481-2440 📠 1-888-RDAVIES

e-mail: rdppub@vir.com

The publisher wishes to take this opportunity
to thank the Canada Council for the Arts,
the BPIDP program at Canadian Heritage
and the Ministère de la culture du Québec (SODEC)
for their generous support for its publishing programs.

Preface

The great age of English aviation wordbooks was exactly 1942 to 1947. An anonymous—author identified only by the initials H W— 15-page pamphlet with simple illustrations called 'What's the Gen?" shows the appearance, from the bottom ranks up, of a lively slang, among RAF and other Allied airmen. It was published by Crowther in London in the third year of the Second World War. At about the same time, for training purposes, the forces produced the Pilot Officer Prune cartoon instruction booklets, which spun off into "boys' books," using genuine slang as she was spoken. Of course, at the end of the war, Eric Partridge's monumental *RAF Slang* (1945, reprinted in 1989 by Pavilion Books and sold out in four years), Fred Hamann's *Air Words*, published in Seattle in 1945/6, and a number of other books made record of the widely known, rambunctious dialect of the aircraftsmen.

It is true that just after the birth of heavier-than-air flight, several books, perhaps more of them in French than in English, had offered, in 1909 and 1910, glossaries of the language of this inspiring, dangerous new pastime of humans (and some wonderful illustrative photographs). And after WWII, the heady confidence in aerial warfare created by the flight of the Enola Gay, the aircraft that delivered the atomic bomb to Japan and won the war in the Pacific, in the popular mind, first gave aviation the prestige to force, in the United States, the creation of the American Air Force, removing air power from the subordinate status to which the Army and Naval Air Forces had been confined. Then came Korea and Viet Nam, which air power seemed not appropriate or adequate to deal with, especially in the latter, guerrilla war. But the appearance of so many aviation lexicons during and just after the Second World War has never been equalled before or since.

The Gulf War of 1991 represented a triumph for air power unprecedented in history, and the flood of words (**CRASS** though they may have been) accompanied the songs of praise.

When we came to compile this book, we could not have defined our goals better than they had been set forth in the preface to Fred Hamann's *Air Words* in 1945:

Aviation is less than a half century old, yet no other industry has originated a language as rich in slang, argot, colloquialisms and colorful terms.

There are two paramount reasons for this phenomenal growth of "air words." First, American flyers, like their countrymen in general, possess a keen sense of the vigor and flexibility of their native speech. Secondly, man's conquest of the air constitutes a new experience for the human race. This combination of circumstances stimulates airmen to invent new words and racy idioms to express their sensations, failures and successes under the impulse of triumphing over the new element.

The language created by airmen today is adding a Buck Rogers flavor to the spoken tongue. This development, of course, is only in its initial phase, for aviation is fast moving into the era of stratoplanes streaking through space at supersonic speeds. Words like *squirt plane*, *blow job*, *jet-propping*, *zizz plane*, *atomplane*, *rocket ship* and even *spaceship* fall with a familiar ring from the lips of flyers and aircraftsmen.

The exciting, vivid language of aviators aroused the interest of the author the first day he became engaged in publicity work for the Boeing Aircraft Company two years ago. Extensive research, innumerable interviews with experienced airmen from every quarter of the globe and the assistance of a score of advisers familiar with the many phases of aviation have gone into the compilation of *Air Words*.

A sprinkling of commonly used technical terms, which everyone who pretends to a knowledge of aviation should know, has been included in the book chiefly to give the reader reference words when seeking synonyms and antonyms. Highly scientific terms have been avoided as being beyond the scope of this work.

The author is aware that his compilation of air words is incomplete, so contributions for future editions of the book are welcome.

He will also be grateful if criticisms of inadequacies in the handbook are communicated to him in care of the publisher.

Seattle, 1946 —F. H.

INTRODUCTION

We see the "Buck Rogers" tendency mentioned by Hamann as having its part in a further expansion of the lexicon which may be described as "every pilot tries to sound like Chuck Yeager." A sort of West Virginia accent, a military conciseness, the prestige of the first man to break the sound barrier—all flavored the increase in numbers as more and more private pilots and the expansion of air travel made the aviation speech community larger and more diverse than it had been when Partridge and Hamann recorded the language of flyers of whom Churchill said "Never in the field of human conflict was so much owed by so many to so few."

In fact, the pilot, at the centre of the linguistic web, is well-known for his extreme individualism in an activity in which he almost always depends on many others—mechanics, weather forecasters, control tower radio operators, and now many computer experts, crew schedulers and trackers, dispatchers, ground crew, and so forth. The stress that comes from being the one human who has to act on all the information tends to produce spectacular relief-of-stress rituals, not the least of which is a habit of speaking that pays little attention to political correctness or taboos. We have tried to reflect all the new forms of usage, at least sampling them; the strengths and limits of this project depend, of course, on the range and experience of the informants and other sources being consulted.

We have gathered material from a variety of aviation and aircraft personnel, including some Air Canada pilots, among them a member of the Gens de l'Air (the Canadian francophone aviation industry pressure group), American Airlines headquarters training personnel; flight attendants; aircraft manufacturing industry executives, engineers, and technical writers (from Canadair, Boeing, Shorts, de Havilland, and Lear); aviation and space writers; such published glossaries as Eric Partridge's 1945 *RAF Slang*, Gulf War glossaries in John and Adele Algeo's "Among the New Words" and a glossary of fighter pilot (mostly U. S.) slang

(*American Speech*). We have done on-site research at air shows in Paris, Berlin, Auburn-Lewiston (Maine), St. Hubert (Québec), Abbotsford BC, and Ottawa. We have scanned books ranging from Chuck Yeager's autobiography and the biography of Grant MacConachie, bush pilot and first president of Canadian Pacific Airlines, to xeroxed fictive accounts of real experience such as Don McVicar's 1991 self-published *The Grass Runway*.

The result to date is this book of lively, evocative, authoritative lingo from the enormous speech community around the world of flight. Keywords in the life of flying, they express the macho, the terror, the care for technical excellence, the struggles over the power of naming between manufacturers' public relations and reporters, between flight crews and ramp rats, between **PAX** (passengers, or "**self-loading cargo**", a few of them **jump-seat sniffers**, as some like to call them) and cabin attendants. They capture and index the sense of freedom and exhilaration from a "**blue on blue**" flying day and also the terror of the "**ground loop**" or a "**torque stall.**"

A number of entries, separated into a section called "Plane Names," are interesting nicknames, from the official names given to models or specific aircraft by manufacturers, military, or airlines, to, often in hilarious dramatic contrast, the folk names given by people who actually had to fly or ride in them (see **Aluminum Overcast, BUFF, Scare Bus, Vomit Comet, Forktailed Doctor Killer,** and **Rosinenbomber** — "raisin bomber" or "currant bomber", the nickname given by Berliners to the aircraft that flew food in during the Berlin Airlift). A definitive guide to plane names is the book by that name published in 1994 by Airlife.

A list of printed sources appears at the end of the text, which may suggest further reading for those interested.

Montréal, 1997. —L.J.P. —M.J.S.

ENTRIES

A. B. C. — Air-borne Cigar, a bomber in the **stream** that carried three German-speaking wireless radio operators who issued orders to the enemy, confusing their aircraft and anti-aircraft forces. WWII. —Doug Sample CD, Canadian Branch, Yorkshire Air Museum.

abeam — directly beside, at right angles to the line of flight. A term borrowed from boat talk.

abnormal positions — RAF WWII term used to describe spins, stalls, inverted flight, out-of-control, etc. as opposed to straight-and-level.

abnormal procedures — used to be called emergencies, commonly applied to engine fires, system failures, etc.

abort — cancel, as for example, a landing or a mission. See **one-oh-nine-itis**.

a/c — in cars this always means "air-conditioning." In plane talk, it means aircraft. Sometimes spelled "ac".

2. capitalized, it means Aircraft Commander; left-seat pilot in a Huey or Chinook, back-seat pilot in a Cobra. —Heath, *CW2*.

ace — a fighter pilot credited with at least five enemy kills. "In June (1916) the French, anxious to divert attention from the carnage on the ground, decided to single out for attention those aviators who has scored at least five confirmed victories in the air. Soon a pilot who met this standard was publicly dubbed an ace; his victory count was published in a running box score in the French press. The Germans eventually adopted the French ace system, but with ten victories as a minimum requirement. They called the flier who reached that plateau 'kanone' or 'top gun.' The British never officially recognized the designation 'ace'; Boom Trenchard, for one, thought it brought undue acclaim to the fighter pilot and diminished the valiant, often sacrificial efforts of the observers and gunners." —Bowen, *Knights of the Air*.

ack-ack — anti-aircraft fire. See also **Archie**, **flak**, **triple A.**

acknowledge — verify that the (radio) message has been received and understood.

ACM — Air combat manoeuvreing, of which the best known form is **dogfighting**. From John and Adele Algeo, "Among the New Words: Gulf War Glossary," *American Speech* (Winter 1991). Another sort of ACM was an Allied WWII formula which had the aircraft attacking out of the sun, making one pass, then leaving. See **beware the Hun in the sun**.

acoustic camouflage — in use on the U.S. **stealth** reconnaissance aircraft (probably unmanned), the employment of sounds 180 degrees out of phrase with the noise made by the aircraft, causing the article to "disappear" from hearing. A form of "adaptive stealth." —Don Hackett.

the **active** — the runway in use. Also **duty runway**.

active camouflage — unconfirmed but likely new technologies used on the U.S. **stealth** reconnaissance aircraft (probably unmanned) include coloration that can change, computer-controlled, to make the aircraft disappear against any colour background, and, again depending on background, lights employed so as to make the aircraft disappear. A form of "adaptive stealth," specifically "visual stealth." —Don Hackett.

active controls — an automatic flight control system wherein vertical acceleration in a sudden gust is counteracted by the upward or downward movement of the controls, useful in **microburst** conditions.

actuals — the description of weather conditions from the most recent observations, the actual weather.

addles — dummy carrier landing decks painted on runways for practice landings.

Adopt-a-Pilot — "Army catchphrase expressing approval of the Air Force because of its role in making the ground campaign easier, competitive raillery being normal between the services." —From John and Adele Algeo, "Among the New Words: Gulf War Glossary," *American Speech* (Winter 1991).

Advanced Aerobatic Cocktail — an aerobatic manoeuvre "for high-performance aircraft. Official sanction is required for exhibition purposes: At 18,000 feet altitude, vertical power-dive to terminal velocity; at about 8000 feet (if wings have not pulled off) make smooth parabola to horizontal level and continue vertical climb; at the "top" (swallow hard) pull over on back and make precision inverted spin of six-and-a-half turns; recover to normal level (cruising) flight. Execute extra large aerobatic figure eight then follow immediately with a snap roll (two turns) to left, precision spin three turns to right and three turns to left; snap roll (two turns) to right; falling leaf (lose 1000 feet); three inside and three outside loops; slow roll to form horizontal figure-eight pattern (upper portion left rolls, lower portion right rolls); inverted falling leaf; inverted approach and normal landing." — Zweng, *Encyclopedic Aviation Dictionary*.

aerial train — one or more gliders towed behind an airplane.

aerobatics — manoeuvres such as loops, rolls, spins, dives, performed alone or in various combinations of aircraft. Also **aeros, stunts**.

Aerodrome of Democracy — what Roosevelt called Canada during WWII.

aerodyne — an archaic term for a power-driven flyer of the aeroplane type, preWWI usage.

aerofoil — British term for airfoil. Also applied to skis or hydroplanes, as in the **pantobase** design.

aerolocomotion — an archaic term for propulsion or powered flight. PreWWI. —Pierce, *Dictionary of Aviation*, 1914.

aeromad — mad or crazy about aeronautics, aeroplanes, or aviation. Afflicted with **flying fever**. PreWWI usage. Hamann records "air crazy" and "air happy." —Pierce, *Dictionary of Aviation*.

Aeromaybe — derisive nickname for Aeromexico.

aeromobile — a vehicle which moves through the air; an automobile of the air; an airship; specifically a flying machine of the helicopter type. PreWWI usage. —Pierce, *Dictionary of Aviation*.

aeroplane — British term for "airplane." According to Bruce Callander, the word preceded the machine by many years; it was, he says, Joseph Pline of France who in 1855 "combined two Greek words" to form the term, which had to compete briefly with "the machine," "the flyer," and "aeronef", just after the Wright brothers actually "made and flew" the "Kitty Hawk Flyer," as the Smithsonian Institute subsequently dubbed it. —"Jargon of the Air," *Air Force Magazine* (October, 1992).

aeros — aerobatic manoeuvres. Also **stunts, aerobatics**.

aerotherapeutics — the treatment of disease by varying the pressure and modifying the composition of the air surrounding the patient. See **whooping cough flights**. —Pierce, *Dictionary of Aviation*.

aeroyacht — an aerial yacht; a fancied airship of the racing or cruising type; a pleasure ship of the air. Pre-WWI usage. —Pierce, *Dictionary of Aviation*.

afterburner — From the German *nachverbrenner*. "engine component which, at the pull of a throttle, begins to burn huge amounts of fuel at high speed, resulting in a great burst of power". — Murray, "....Navy Fighter Pilots." Essentially, an afterburner dumps fuel into the exhaust of a jet engine. According to John Wheeler of Boeing, "afterburners may be used even on takeoff (as is done on the Concorde, for example) for additional thrust." Brits know it as **reheat** and **re-light**.

Agony Airlines — derogatory name for Allegheny Airlines.

aileron — from French, literally "little wing." A minor wing control surface, introduced after **wingwarping** had been used, to help stabilize and control the aircraft.

J. A. Foster in *Sea Wings* tells a story of the origin: "In September at the Coney Island race track when McCurdy explained to French airman Maurice Farman how the new wing flaps operated, Farman's eyes lit up. 'Ah,' he exclaimed, 'Ailerons!' His name for the 'little wings' stuck." C. 1910? Gordon Baxter in *More Bax Seats* (1988) suggests whimsically, "what Glen Curtiss had to invent to avoid flying a Wright."

aileron roll — See **roll**.

Air America — pseudo-civilian company operated by the CIA.

Air Apparent — facetious nickname for Air Canada, according to John Cavill, public relations rep for the airline's eastern Canadian region in the 1980s.

airboy — in the early days of aviation, young boys were often employed to help passengers aboard, load the luggage, etc. Typical of early Luft Hansa (form of the airline's name prior to WWII) operations.

airbrakes — see **dive brakes**, **speedbrakes**.

airbridge — British term for the closed walkway connecting the terminal departure area and the aircraft, for passenger embarkation. In North America usually called the **jetway**.

Air Canada captains — according to Joyce Spring, in *Daring Lady Flyers*, about the early achievements of women in Canadian aviation, Air Canada captains were "overconfident, oversexed, and overpaid."

air carrier — an air line, cargo or passenger, scheduled, nonscheduled, or charter.

Air Chance — facetious nickname, in the somewhat distant past, for Air France. An explanation for the origin of this name is offered by John McPhee, in a profile of Temple Fielding, of the Fielding Guides to Europe: on a flight from Copenhagen to Paris, he stood beside the pilot, who smoked while the aircraft was being refueled; then watched in horror as the pilot flew to Paris at 1000 feet, during which time the flight attendant carried trays of cognac into the cockpit. Fielding went public with the nickname, causing Air France to ban consumption of alcohol in-air by flight crews in 1964. ("Temple Fielding," originally published in the *New Yorker*, was reprinted in McPhee's book *A Roomful of Hovings*).

air circus — alternate description for **flying circus**, as in Baron von Richthofen's infamous formations of WWI. The term has since been applied to any mass or formation of aircraft.

aircraft — see **airplane**.

aircraft carrier — this well-known name for a ship with a flat deck from which aircraft may be launched has entered basketball slang as a phrase for a big, skillful player. Many terms in b-ball (like Air Jordan, and the movie "The Air Up There") refer to air and sky, because the ball has to get up there and come down in the right place.

airedale — British nickname for naval aviators.

airfile — term used when filing a flight plan while airborne, almost always on Instrument Flight Rules. See **pop-up IFR**. Also see **fast file**.

airfoil — 1) wing shape that provides lift. 2) crew captures hijackers — whimsical definition by Gordon Baxter in *More Bax Seats* (1988). 3. Also whimsical, sword used for in-flight dueling.

airframe — the actual body of the aircraft, including the wings, fuselage, tail assembly, etc., as distinguished from the engines, electronics and avionics.

AirLand Battle doctrine — "Use of synchronized tactical air and ground assault in a rapid, massive penetration into enemy territory, in opposition to traditional massed opposing armies; developed for application against the Warsaw Pact forces but applied in the Gulf War, the fusion of the term "AirLand" symbolizing the combination of ground and air power." —From John and Adele Algeo, "Among the New Words: Gulf War Glossary," *American Speech* (Winter 1991).

airlifter — generic term for any heavy cargo transport, such as the

Lockheed C-5 Galaxy or the McDonnell-Douglas C-17.

airline pilots oral rules — 1. get the killer item (i.e. fog is not a killer item, mountains are) —David Hoover, American Airlines pilot and one of two builders and fliers of the Coors Silver Bullet (a one-man 324 mph jet like James Bond's in the movie "Moonraker" and more recently known as the BD5-J).

2. Keep the shiny side up and the greasy side down. (Also a trucker's rule).

3. Joke rule: "pull back if you want to go up; if you want to go down, pull farther back." (sarcastic rule to test recruit [if he writes it down and seems to believe it, flunk him]): a variation: "take it/fly it low and slow, and throttle back in the turns." "I never pull back when flying low, so as not to drag the tail wheel!" —Pilot Officer Prune, Raff, 1942. See **low flying.**

4. Service, Schedule, Safety — Air Canada pilots' rule of thumb. A former flight attendant remembers the order as "Schedule, Service, Safety." (It is perhaps unfair to point out that here's one case where safety isn't First, because no doubt pilots, like sailors, can do more than one thing at once!)

5. "There are old pilots and there are bold pilots but there are no old bold pilots."

See also **bush pilots' oral rules, military pilots' oral rules.**

air mail — "Mails may be lost, but must never be delayed; and passengers may be delayed, but must never be lost." —C. G. Grey, ed. *The Aeroplane,* 1937.

airmiss — British term for **midair**, possibly shortened from "air mishap."

airplane — synonyms: **aircraft, a/c, kite, bird, ship, crate, bus,** cab, equipment, **aeroplane** (Brit.) Millions of spare parts and miles of wiring flying in close formation. According to Ben Rich, of Lockheed **skunkworks**, "airplanes are like people. They tend to gain weight as they get older. The first time the U-2 took off to overfly Russia back in 1955, it was a svelte youngster at 17,000 pounds. Now it had ballooned in middle age to 40 percent over the original model and bent the scales at 40,000 pounds." *Skunk Works.* Another way of seeing how aviation people see the machines as "alive" is in the realization that types or designs of aircraft tend to evolve from earlier ones rather than being totally new: the de Havilland Dash-8 was a good aircraft to replace the ATR when the latter were grounded for icing problems because Canadian aircraft were always overbuilt, to cope with, for example, icing problems.

airplane driver — pilot (U.S. Air Force use) ("Among the New Words," Gulf War Words supplement, *American Speech*, 1992).

airplane spruce — "Sitka spruce (*Picea sitchensis*) of the wet coast was cut during both world wars for use in aircraft construction. The wood has straight grain, rarely twists on drying, and has a high strength-to-weight ratio. These characteristics are ideal for building **airframes**". —Tom Parkin, *WetCoast Words*.

airpocket — "An airpocket is something which only newspaper men know of because there has never been a flyer who has found one. There is some kind of myth or legend among newspaper men about a space in the air in which there is no air, therefore it is unable to support an airplane when it runs into it. Anyone who knows anything about the action of air knows that no such place can exist." Xelphin V, Dugal in *Readings on Aviation* (1929). Despite this denial, Fred Hamann (1945) records continuing belief in their existence in an entry for "air holes: descending air currents that cause the plane to drop suddenly as though it were falling into a hole." Sometimes called **bumps, chop, turbulence**.

airports — "In the past fifty years airports have undergone major transformations. They have developed from aerodrome to airport. From public utility to commercial entity. From administration of a simple function to management of a complex enterprise. From providers of services to airlines to enterprises marketing to **PAX**. From primary facility to intermodal transportation centre, indeed a platform for many commercial activities and a partner in economic development." —Phillippe Hamon, director general of Airports Council International (ACI Europe) in *Financial Times*, (November 15, 1994).

air rep — same as **pirep**, a report on flying conditions from aircraft in the air.

airscrew — propeller, as originally named by British flyers, but then replaced by the American "prop" or "propeller" during WWII. (*Time* magazine, March 22, 1943) A story goes that an RAF engineering officer during WWII had a problem whenever he requisitioned an airscrew; he'd receive instead, an air crew, ready to take out a flight. To avoid misunderstandings of this type, the Air Ministry declared that henceforth the term 'propeller' would be used. —from *Pilot Officer Prune's Picture Parade* (1942).

airship — a lighter-than-air vehicle, e.g. blimp, dirigible, Zeppelin.

airspeed, air speed — comes in several forms: IAS (indicated air speed—what the gauge says), CAS (calibrated air speed - gauge reading corrected for drift, etc.), and EAS (equivalent air speed - includes factors such as density; airplanes always stall at the same EAS for given weight and same **angle of attack**). Speed at which the aircraft is moving through the surrounding air, not necessarily **ground speed** or **true speed**.

airstairs — self-contained stairs or ladder that fold up and into, and become part of the aircraft.

airstart — starting an aircraft engine while in flight, especially a jet engine after a **flameout**.

airstrip — synonym for airport. Also, jocularly, an in-flight performance by an exotic stewardess.

air-to-mud — aerial delivery of weapons on ground targets in support of ground troops.

airtray — "a protective covering that fits around a casket. Made of heavy duty cardboard." —American Airlines training materials.

airways — "three-dimensional 'air highways' located at upper altitudes and used by commercial aircraft." —Borins, *The Language of the Skies*. Air routes designated and controlled by air traffic control authorities and provided with radio navigation aids.

Airwing Alpo — "meals ready to eat," ("Among the New Words," Gulf War Words supplement, *American Speech*, 1992).

airworthy — an aircraft's condition of being mechanically sound and safe to fly. Specifically, this term can be applied to any aircraft at any time, and is sometimes expanded to include aircrew. See **MEL**.

Alfa — code word for the letter "A" in the new set (replacing the old Able Baker Charlie, etc.); many uses; one is in **high Alfa** (where "A" means "angle of attack" or a nose-up **attitude**. Note spelling.

alfa damage — damage which makes an aircraft of no further use except for the possible salvage of parts.

Alice Blue Gown — fine weather, clear skies. "I'm going up Alice Blue Gown."

Alitalia — Jocularly, "the airline with hair under its wings." —from Robert MacNeil. See also **All Landing in Tokyo, All Luggage in Amsterdam, Always Late in Takeoff, Always Late in Arriving.**

Allied code names — as morale boosters, or propaganda, or because they did not know what the enemy called their planes, the Allies gave a code name to each enemy plane type to differentiate them. The list of Japanese aircraft, for example, includes Tojo (Nakajima Ki 44 Shoki - Demon), Zeke (Mitsubishi A6M Reisen - Zero), Frank (Nakajima Ki 84 Hayate - Gale), Jack (Mitsubishi J2M Raiden - Thunderbolt), Tony (Kawasaki Ki 61 Hien - Swallow), Nick (Kawasaki Ki 45 Toryo - Dragon Killer), Irving (Nakajima J1N Gekko - Moonlight), Oscar (Nakajima Ki 43 Hayabusa - Peregrine Falcon, orig. Jim), and Nate (Nakajima Ki 27 - orig. Abdul but Nate from 1943).

alligator — nickname for "navigator," (rhyming slang) in WWII. —McIntosh.

Allison V-12 — engine used on rail jobs in the top fuel dragster class, AA Unlimited, and also to power the P-51 Mustang fighter aircraft, P-38, and Unlimited hydroplanes. (The P-51 was even more successfully powered by a Packard Merlin, from Rolls Royce Merlin built under license). In Formula I racing, the cars were lighter but still had enormous (three to six litre capacity, up to 24 cylinder) engines.

All Landing in Tokyo, All Luggage in Amsterdam — derogatory explanation of the letters in Alitalia, the Italian national airline. See also **Always Late....**

all three dead — phrase said to explain the meaning of "A3D" in the U. S. Navy's Douglas A3D aircraft, because the naval version has no ejection seats. Also known as a **whale.**

all up weight — total weight of an aircraft at take-off. WWII.

Allergic to Combat — see **ATC.**

Alpha strike — A large offensive air strike, involving all a carrier wing's assets, fighters, attack, refueling, etc. —Mersky & Polmar, *The Naval Air War in Viet Nam.* See also **MITO, flush launch.**

altimeter setting — Jocularly, "the place where the altimeter sets....usually hidden behind the control column during a tight instrument approach." —Bob Stevens, *"There I was."*

altitude is life insurance — WWII pilots' saying. "Let's buy some life insurance," that is, climb to a higher altitude.

altiports — airports constructed at high altitudes, such as in the French Alps during the 1960s.

aluminum clouds — "On many occasions they [the Russians] were flying squadrons fifteen thousand feet underneath the U-2, trying to block the view. Kelly Johnson called that 'aluminum clouds.'" —Ben Rich, *Skunk Works.*

aluminum overcast — a crowded sky. See also "Plane Names."

Always Late in Takeoff, Always Late in Arriving — derogatory explanation of the letters in Alitalia. See also **All Landing**.....

Another Useless Airline — irreverent nickname for Austrian Airlines (AUA).

anchor — an order to "orbit about a specific point; ground track flown by tanker....a turning engagement about a specific location." —from Canadian Forces Aircom Regulation Glossary.

angels — altitude in thousands of feet as in "angels 24", meaning 24,000 feet; a term popular in the Korean War era but seldom used today, though it is listed in a contemporary Canadian Forces Aircom Regulation Glossary. Measurements in hundreds of feet are called **cherubs**. The term "flight level" is commonly used today. 2. **SAR** helicopters or aircraft.

angle of attack — see **attitude**.

Anti-Surface-Vessels — (ASV) British WWII term applied to radar-equipped aircraft, especially Fairey Swordfish, assigned to hunt and destroy enemy shipping.

antlers — type of control column wherein the wheel is crescent shaped, like horns or anters. 2. early style of nose-mounted radar antennae, WWII. See **stag's antlers**.

Antonov — USSR design bureau aircraft manufacturer, headed by Oleg K. Antonov, Kiev, with aircraft firm headquarters in Minsk, Ukraine, which built the Russian aircraft known by that name. Among his most famous designs are the An-2 **Colt**, a bulky biplane first built in 1947. "It appeared to be an obsolete mistake. So unmistaken was it that it has been manufactured in larger quantities than any other single type of aircraft since WWII. Soviet production topped 5000." —Bill Gunston, *Modern Soviet Aircraft.*

Similarly, the MiG was built by the design bureau of Mikoyan and Gureyevitch, and the Tupolov by a bureau headed by someone of that name. The last-named was known for close copies of the 727 (the Tu-134) and of the B-29 (the TU 4).

apron — a hangar's **tarmac** surround. See also **ramp, pan**.

aquaplane (v.) — condition during which landing aircraft is travelling at high speed through shallow water and is entirely supported by dynamic reaction of the water, tires out of ground contact, making steering or braking difficult to impossible.

Archie — anti-aircraft fire, from WWI. See **ack-ack, flak**. "To Allied fliers, antiaircraft fire was 'Archie,' so named by a British pilot who encouraged himself whenever a shellburst rocked his plane by shouting, 'Archibald, certainly not!' —which was a London music-hall refrain sung as the show's leading lady fended off a lecherous suitor." —Bowen, *Knights of the Air*. Also **triple A**.

area rule, area ruling, area ruled — a tenet, a philosophy, a process, by which, for example, the design of a fuselage is given a pinched waist for better airflow at supersonic speeds. Known as the **Coke bottle** or **Marilyn Monroe** design. "I did the area ruling on the Challenger." "That's been area ruled." The term refers to the predictive measuring of the air resistance of the fuselage surface.

armchair flyer — one who talks flying without actually doing it.

Army of Terrified Co-pilots — see **ATC**.

arresting gear — also known as the **hook**, wires strung across a carrier's deck, to snag a hook extended below a landing aircraft's fuselage; landings thus accomplished are called "arrested landings."

arrow — "fighter plane that flies in a straight line for an extended period of time, with no **jinking** or **yanking and banking**." —Murray, "Language of Naval Fighter Pilots."

artichoke — code name for **gardening** operations off the coast of France, in mine-laying operations from the air during WWII. **Sweetpea** meant in the Baltic; **nectarine**, off Holland.

article — the airplane being tested. "Article was stable in taxi mode." Often referred to as "test article."

ashcan — gun emplacement turret on a WWI military aircraft. Also known as a **trashcan** or **dustbin**.

Aspirin — see **dog-leg, headache**.

ass — U-2 pilots "needed an iron butt for ten-hour flights. 'I ran out of ass before I ran out of gas,' some U-2 **drivers** would complain." —Ben Rich, *Skunk Works*. In another example, "Refueling could extend the U-2's range to seven thousand nautical miles and fourteen straight hours of flying time, which pushed a pilot's fatigue beyond safety....'Never again. My mind went numb ten minutes ahead of my ass.'" —Bill Park, quoted in Rich, *Skunk Works*.

Atagirls — Enlisted women pilots in the British Air Transport Auxiliary, WWII. The women were needed because there "were not enough unfit civilian male pilots ("ancient tattered airmen") to ferry fighter planes and bombers to military bases." —"Social Studies," Toronto *Globe and Mail*, June 9, 1994. ATA was said to stand for "Anything to Anywhere." —Joyce Spring, *Daring Lady Flyers*. ATA was said to stand for "Anything to Anywhere."

ATC — 1. Air Traffic Control. 2. WWI, Air Transport Command. 3. **Allergic to Combat, Army of Terrified Copilots.** —Leuthner and Jensen, quoting Len Morgan.

atmospherics — static, as heard on radio transmissions.

your **attitude determines your altitude** — complex proverb or rule of thumb which seems to derive from the scientific truth that the aircraft's "attitude," or alignment in relation to ground and wind currents (nose up, nose down, climbing, descending, banking or straight and level, etc.) is related intimately to safety, position, and success in attaining flight objectives. Metaphorically, it seems to refer to presence of mind, calm, balance....

atoll — "direct hit, from the name for air-to-air missiles that Russian-built MiG-21s carry." —Murray, "....Navy Fighter Pilots".

auger in — drill a hole in the ground. **Buy the farm**. A reference to the spiralling motion of an airplane spinning out of control out of the skies and crashing nose first. See **smoking hole**.

auntie — (USAF) anti-, as in "antimissile missile." —Paul Dickson, *Slang!*

autocat — an aircraft equipped to serve as an airborne communications relay.

autopilot — an automatic device that actually does most of the flying, most of the time. See **Mile-High Club, George..**

autorotation — engine-out procedure in a helicopter, controlled descent via windmilling blades.

available — not quite ready to fly aircraft: "might mean anything from needing a tyre inflated to needing a major overhaul." —Deighton, *Fighter*. See also **serviceable**.

avchat — what you are reading now, according to Wilson Leach, *Aviation International News*. Another term for "plane talk."

AVG — American Volunteer Group, the official name of Gen. Claire Chennault's Flying Tigers. One squadron was named "Hell's Angels." (Howard Hughes, who built the **Spruce Goose,** largest plane of its time and for many years, made a movie in the '30s about the Tigers named "Hell's Angels.")

avgas — aviation gasoline.

aviatrix, aviatress, aviatrice — various terms in use in the past for "female flying enthusiasts," with one citation to the OED listing an entry in the Daily Chronicle of 1910. —"Social Studies," Toronto *Globe and Mail,* June 1994.

Avianunca — derisive popular nickname, in Spanish, for Avianca, the Colombian national carrier, which means "never arrives," or literally "Aeronever." ("Taking to the skies can be adventure in absurdity," *Toronto Globe and Mail,* July 4, 1994).

avionics — navigation systems, using the flight instruments, particularly applied to later-generation systems like radar, VOR, DME, etc.

B-lock — boredom-induced loss of consciousness which may afflict transport pilots. David Hoover, stunt flyer and American Airlines captain (see **airline pilot's oral rules**), routinely answered "what do you do?" with "I drive a truck," obviously thinking of his day job ferrying passengers rather than **stunting** the Coors Silver Bullet. Also spelled B-loc. See **long hours....** and also **G-lock.**

baby barrier — In 1993-94 several extremely young pilots made the headlines for being youngest transcontinental **PIC** or youngest transoceanic PIC, etc. But what's the very youngest one can fly? According to USAF Flight Surgeon Dr. Richard Niemtzow, "I suspect there's a threshold, below the age of eight, where there are limiting physical and cognitive functions." He added it depends on the child's level of maturity and ability to grasp complex concepts. "Certainly, below the age of seven would be a cutoff. My son started flying when he was ten-and-a-half, and there, the limiting factor was whether he could sit high enough in the seat to see over the panel and still reach the rudder pedals." Nine-year-old Rachel Carter, who set several distance records in a Piper Arrow, says the baby barrier is at around eight years old. "Any younger and a lot of moms wouldn't let their kids do it. You have to know math, be able to read, know English and some social studies." —- Phil Scott in *Private Pilot Magazine,* July 1994. See **bi-**

ology barrier, thermal thicket, sound barrier.

back-room boys — British term for **boffins**, scientists or aeronautical engineers of the WWII era.

backseater — "radar intercept officer, who sits directly behind the pilot in the cockpit of the fighter plane." —Murray, "Language of Naval Fighter Pilots." Compare **gib, gibs.**

backside of a power curve, behind the power curve — Too much drag, not enough speed. "Adopted from a graph of an aircraft engine's RPM relationship to horsepower output, which declines suddenly beyond design limitations. A pilot may be said to be flying the backside of a power curve if he has overloaded his plane, or if he's 'bending an elbow' in the local pub." —Tom Parkin, *WetCoast Words*. Parkin's gloss points to metaphoric adaptations of this very key phrase in early flying. Essentially it referred to any situation in which more power was or might be suddenly needed than was available. For example, with any aircraft, a way to gain speed was sometimes to dive. However, if one dived too low, there might not be enough power available from the engines to pull out, get the nose up enough to gain needed altitude. So you'd be **"behind the power curve"** or on the "backside of a power curve," and maybe out of luck. —from Bob Parke.

Also **behind the airplane,** meaning "he/she is not keeping up with control and power in proportions needed to fly this aircraft." —from Paul Turk.

backside of the clock, flying the — night flying. Many **freight dogs**, or cargo pilots, fly at night (according to Barry Lopez, some "wear bat wings instead of eagle wings.")

back stick — pulling back on the **joy stick** to make the nose rise.

back-to-back — 1. "flight manoeuvre in which two fighter planes position themselves one over the other, with the top one upside down." —Murray, "....Navy Fighter Pilots".

2. "Two reduced-fare round-trip airline tickets used to avoid the expense of a single higher-fare ticket (e.g. a traveler wanting to go from Atlanta to Chicago on Tuesday and return on Wednesday—a high-fare midweek flight—buys one round-trip ticket leaving Atlanta on Tuesday and returning the following week—and another round-trip leaving Chicago on Wednesday and returning the next week, and uses only the first half of each ticket). See also **split ticket, looped, nested.** —John and Adele

Algeo, "Among the New Words," *American Speech*, Spring 1994.

Bader's Bus Company — "The hottest sector of the southeast was immediately east of Portsmouth at Tangmere, where a three-squadron wing of Spitfires, led by Wing Commander Douglas Bader, the legless ace of the Battle of Britain, was headquartered. The wing comprised Squadrons 145, 610, and 616, and was affectionately known to the British and distastefully to the Germans as Bader's Bus Company." Munro, *The Sky's No Limit*. WWII. Bader had lost both legs in a pre-war air crash, in the late '30s, but he came back to be a Spitfire ace.

Baedeker raids — German bombing raids on such historic towns and cities as Exeter, Bath, Norwich, and York during 1942, believed to have been made in retaliation for Allied bombing raids on such German cities as Cologne.

bag throwers — irreverent nickname for baggage handlers. Also **baggage-smashers**.

bakery — translated from the German, Lufthansa's slang term for its pilot training programme. It was said that the trainees were being "baked." —from Grant McLaren's article in *Professional Pilot* (Summer, 1992).

bail out — parachute to save one's life. See **Caterpillar Club, Walker's Club**. Often spelled bale out. See **eject**.

balbo — "a large formation, a crowd; named after Air Marshal (Italo) Balbo, who led twenty-four Italian aircraft on their Atlantic flight." —W. H. *"What's the Gen?"*.

ball or **meatball** — an optical landing system for landing on an aircraft carrier. "This device used a yellow light arranged between two green reference, or datum, lights to give the pilot a visual indication of his position in relation to the proper glide path. If he kept the ball centered in the datum lights all the way to touchdown, he would catch the third of four arresting-gear wires rigged across the deck." —Coonts, *Flight of the Intruder*.

balloon busting — Royal Flying Corps phrase from WWI meaning "to attack enemy observation balloons," which were also known as **sausages, pigs**.

ballooning — a term applied to the upward sweep of an airplane when a pilot, misjudging his landing, pulls back on the stick just before the landing gear touches the runway.

balls — Low flying. "We were coming in over those darned pines, and I reached down and caught my balls and held them up so they wouldn't hit on the tree tops!" —RCAF instructor, BCATP quoted in *The Plan*. See **contour flying, counting the insulators, nap-of-the-earth**..

balls out — "giving it all you've got." —from Chuck Yeager's autobiography.

ball-trap — "an at times unexpectedly collapsible seat, specially in an aircraft/RAF: since 1940. Since 1945 Australian civilian for tractor-seats, etc." —Partridge, *Dictionary of Slang and Unconventional English*, 1989.

ball turret — spherical shaped machine gun enclosure, with 360-degree rotation, usually located on top of fuselage or under the belly, as in the Boeing B-17 Flying Fortress. Sometimes called a **bathtub**.

From my mother's sleep I fell into the State,
And I hunched in its belly till my wet fur froze.
Six miles from earth, loosed from its dream of life,
I woke to black flak and the nightmare fighters.
When I died they washed me out of the turret with a hose.

—Randall Jarrell, "The Death of the Ball Turret Gunner." (postWWI).

banana peelings — "cowling between the cylinders." —Howard N. Rose, *A Thesaurus of Slang*.

bandit — 1. **bogey**. However, "in RAF, any enemy aircraft as opposed to a Bogey, which was an unidentified subject." —Group Captain J. A. Kent, *One of the Few*. 2. unlicensed or improperly licensed air carrier or taxi.

bank — tilt the aircraft to right or left while turning.

bang out — "eject from an aircraft" (UK source) — From John and Adele Algeo, "Among the New Words: Gulf War Glossary," *American Speech* (Winter 1991). See **bail out**.

barbette — an external pod, a cylinder of armour, containing and protecting weaponry such as a machine gun or cannon, usually mounted underwing or in the tail of a **warbird**. Usually an "add-on," rather than a structural, integral part of the fuselage like a canopy or a turret. From the French (1772), "a mound of earth or a protected platform from which guns fire over a parapet" (OED), from the French for "little beard."

BARCAP — Barrier Combat Air Patrol, a fighter patrol between the carrier task force and enemy threat.

—Mersky & Polmar, *The Naval Air War in Viet Nam*. See **SARCAP**.

barf bag — airsickness bag, provided by airlines in the pocket with emergency instructions, location of exits, etc.

barf hop — sightseeing excursion flight. See also **flightseeing, hick hop**.

barnstorming — an old name for stunt flying, presumably because in early flying the available landing spaces were mostly fields, and people would gather near barns to watch the **aerobatics**. The stunt pilot would **buzz** the barn, to give them a thrill. And the hangar is even now known occasionally as the "barn." "Roll her into the barn for the night!"

Bruce Callander argues that the term had already been used "for decades by traveling theatrical troupes." But he acknowledges that the travelling flyers "hangared their Jennies in barns" and "staged rural air shows." —"Jargon of the Skies," *Air Force Magazine* (October, 1992). See **stunt, wing walking, hick hop**.

barrel — fuselage or fuselage section, also known as a **plug**.

barrel roll — an aerobatic manoeuvre in which the pilot does a combination of looping, banking, and rolling, following the path that would be taken by a spring, or, visually, "flying around the barrel." "An open positive 'G' aileron roll." For a somewhat different definition see **roll**.

base leg — 1. the approach leg flown perpendicular to the runway.

2. stewardess with a deep voice — whimsical definition by Gordon Baxter in *More Bax Seats* (1988). See **downwind leg** and **circuit**.

basic — short military slang for basic training. **Sprogs** use it to mean basic flying training.

basket — term for the **drogue** portion of an air-to-air refuelling pod. "The US Navy uses a female system on its aerial refueling aircraft (tankers). These aircraft extend (from refueling pods) flexible hoses with drogues (called baskets) at one end. These provide a target for the pilots of the aircraft to be refueled. The US Air Force, on the other hand, uses a male refueling system on its tankers. These attach to a refueling receptacle on the aircraft to be refueled. USAF tankers (usually a KC-135 or KC-10) can be modified with hoses and baskets to refuel Navy planes, however," — Eric Hehs.

basketwork — a geodetic design principle for airframes typical to British WWII bombers, such as the

1938 Wellington, such construction being popularly known as "basketwork."

bathtub — "Protective shield against ground fire in an A-10 [slow, heavily protected battlefield-support jet plane]" — From John and Adele Algeo, "Among the New Words: Gulf War Glossary," *American Speech* (Winter 1991). In the A-10 the "bathtub" is made of titanium. Bruce Callander reports that the ball turret was known to gunners who worked in it as "the bathtub".

batsman — carrier landing officer who directs landing aircraft with a pair of ping-pong-like paddles. Also, a ramp person who directs aircraft into its parking bay or space using circular, often lighted, bats. Also **marshaller**.

battlesky — the scene of an aerial battle, according to Arthur Thurston, *Bluenose Spitfires*.

Bavarian jet pilot — men from Bavaria were prized as test/jet pilots because they had no neck and were short and chubby, so "their heads weren't so far from their hearts and the blood didn't have so far to travel" in cases where **G-lock** was a danger, when G-force might be so great as to cause **blackout**. —from Peter Otto, who admits he was one.

beanie —BN, bombardier/navigator, right-seat rider in the A-6 Intruder. —Coonts, *Flight of the Intruder*.

beard — radiator or air intake slung below the engine. Not to be confused with **moustaches**.

beat up — **buzz**, or make low-level stunts and manoeuvres. "To fly over and beat up the field."

beating up the girl friend — stunting over the girl friend's house, one of the earliest forms of aeroexhibitionism.

beaver tail — flat-planed tail surfaces for better air flow.

beehive — WWII term for any large mass of aircraft.

beer wagon — the squadron's utility aircraft. RAF 601 Squadron used a captured Junkers Ju-87 Stuka.

Beethoven — German WWII code name for **Mistletoe** program, a system for ferrying a fighter part way on its mission, mounted "piggyback" on a bomber, to save fuel. See **Father and Son**.

beetle's eye — optically flat clear panels forming the nose section, such as in the Junkers Ju-88.

behind the airplane and **behind the power curve** — see **backside of the power curve**.

bellhop — a specially equipped aircraft serving as a radar or radio relay.

belly-for-rent — what Brazilians call the Cessna Caravan because of its cargo pannier. —from *Business and Commercial Aviation* magazine.

belly landing — a wheels-up landing. See also **pancake in**.

belly — the lower hold or compartments under the main deck of an aircraft in which cargo is stowed. The terms "belly capacity" and "belly cargo" refer to the lower hold, which is sometimes called the "belly pit," or **pit**.

below bingo — USAF radio code indicating a low fuel situation. See **bingo**.

bent his sword — caused damage to aircraft. —Gord Squires, "Joy Riding in an F-86", *Aircraft* (April-May 1990).

bent plane — damaged aircraft. Confirmed by an entry in From John and Adele Algeo, "Among the New Words: Gulf War Glossary," *American Speech* (Winter 1991) — "'bent' (of equipment) inoperative, temporarily or permanently."

bent wing — a **gull** or **inverted gull** wing. Also **cranked wing**.

bereavement fare — "an airline discount fare for those who make a late reservation because of the death of a relative." —John and Adele Algeo, "Among the New Words," *American Speech*, Spring 1994.

beta range — reverse pitch propeller setting, "alpha" being traditionally the opposite.

Better Walk If Able, But Will I Arrive? — derogatory explanations of the acronym BWIA, officially British West Indies Airline.

beware the Hun in the sun — British and American WWII (possibly WWI) fighter pilot's axiom reminding him that German aircraft were prone to attack directly out of the sun, especially in the morning when they had the glare behind them as they came from the east. See **fighter pilot's oral rules**.

Big City — what RAF bomber crews called Berlin. —Spencer and Harris, *Reap the Whirlwind*.

big friends — bombers, especially as referred to by fighter pilots..."the big friends flew a staggered box formation, their guns unlimbered and poking into the slipstream." —Robert S. Johnson, *Thunderbolt*. See **bomber's little friends, little friends**.

Bingo — In "This is Red Two, Bingo," means "minimum fuel for return to base." —Bob Stevens, *"There I was."* See also **Joker**.

Among "stewardesses," though, "bingo" means "a major air crash," according to John Davis, *Buzzwords*. See also **French kiss**.

biology barrier — as opposed to the "sound barrier," this term denotes the problem incurred when the aircraft will go faster than the crew can physically bear. For some time now, some equipment can perform at, say, 9G (nine times the force of gravity) without damage, but the pilots will black out after 90 seconds. See **sound barrier, thermal thicket, baby barrier**.

biplane, bi-plane — early aircraft planform design incorporating two wings, one above the other; "So many wires and struts that engineers said 'if a bird could fly out, you'd forgot one.'"

bird — "helicopter that drops parachutists (a specialization of the general sense 'aircraft'")— From John and Adele Algeo, "Among the New Words: Gulf War Glossary," *American Speech* (Winter 1991). Also, a slang term for any **aircraft**.

bird cage — synonym for biplane because of all the wires and struts. See **boxkite** and **Stringbag** (Plane Names section).

bird-dog — in fire-fighting, a small plane (e.g. Cessna O-1, high-wing all-metal airplane) which, using a siren so it can be followed in the smoke, leads a big water bomber like a Martin Mars into the fire. **Forward fire control** aircraft. Also used to designate a fire-spotter aircraft. See **water bomber**.

birdman — (USAF) pilot. — Paul Dickson, *Slang!* Common term for aviator prior to WWI.

birdpeople, birdperson — politically correct form for **birdman, aviator, fly-boy, airman**.

bird repellent — antifreeze, which is applied to the wings, in jocular naming.

bird shot — an aircraft manufacturers' testing procedure wherein whole (dead, frozen, then thawed) chickens are fired at windshields, engines, nacelles, leading edges, etc. to determine the component's ability to withstand **bird strike**.

bird strike — term used to describe the striking of birds by aircraft on take-off, landing or in-flight, and the damage thereby caused.

biscuits — the three-piece mattress used on RAF cots during WWII. —Doug Sample CD, Canadian Branch, Yorkshire Air Museum.

biscuit bomber — an aircraft that drops food and supplies to ground troops (as long ago as Burma, WWII). Compare **Rosinenbomber**.

bit the biscuit — was killed. See also "went for a **Burton**," "got the chop", **auger in, buy the farm**.

bizjet — a business jet, also known as an **exec-jet**.

a **black** — see **put up a black**.

black box — the container(s) for the automatic recording of flight data and cockpit conversation, useful after crashes in reconstructing the reasons for the disaster. Sometimes there are two: a flight data recorder and a cockpit conversation recorder, usually painted a bright colour, seldom black.

2. common jargon for any piece of avionics equipment. According to one source, in WWII, with the development of increasingly complicated electronic devices, it is said that "frustrated by the technicalities of a new sort of air war he could not master, Göring condemned them all as unnecessary 'black boxes.'" —Deighton, *Fighter*.

"Why don't they make the whole airplane out of the same indestructible material?" — unidentified TV comedian.

black gang — the mechanics. The term seems to have been coined by Alan Bill, a reporter for the Winnipeg *Tribune*, in reporting on the search and rescue of the MacAlpine Expedition in the Canadian Arctic, 1929. See **troops, black men**.

black hats — in parachuting, the jump trainers, at, e. g. Fort Benning, GA. —Pat Dudley. See also **red hats, riggers**.

Black Hole — an infrared signature suppression system, such as on the Hughes 500M Defender helicopter, used in a covert role by U. S. Army Special Forces. —*The Naval Air War in Viet Nam*.

black men — German term for the aircraft mechanics and ground crew because of their black overalls. WWII. —Deighton, *Fighter*. **black gang, troops**.

blackout — loss of consciousness due to **G-force**, lack of oxygen, injury or other factors. See **brownout, grayout, redout, whiteout**.

blackouts — dark-coloured WAAF-issue knickers, winter weight (WWII). See also **twilights**. —Doug Sample CD, Canadian Branch, Yorkshire Air Museum.

blast fence — a barrier near the take-off end of a runway, or in a run-up area to divert engine blast.

blast pens — revetments into which aircraft were placed to protect them from enemy bombing and strafing attacks. WWII. —Deighton, *Fighter*.

blemished with blisters — said of the Tupolev Tu-16 Badger, because the wheel wells, radar installations, and visual observation stations were made possible by **blisters**.

blind flying — flying on instruments, with visibility virtually at zero. See **under the hood, hoodwork, on the gauges, on the clocks**.

bliss — "escape and evasion technique taught to pilots, from 'blend, low silhouette, irregular shape, small, secluded'" (UK source) — From John and Adele Algeo, "Among the New Words: Gulf War Glossary," *American Speech* (Winter 1991).

blister — plexiglas dome, often twinned, one on each side or top and bottom or tail and nose on such craft as the Consolidated PBY Catalina, the Boeing B-29 Super Fortress, and the North American B-25 Mitchell bomber, to give greater visibility. See **blemished with blisters**.

blitz — from **blitzkrieg,** German word for "lightning war," rapidly a new English word for the bombing of London. "Blitzkrieg" became widespread in use after the dramatic success of Germany's invasions of other parts of Europe was accomplished by the combination of air attack (bombing, strafing) and swift movements of tanks and other ground forces.

Jean Kinloch, annotating HW's *What's the Gen?*, a 1942 pamphlet of RAF cartoons and slang, noted that "a solid lump of blitz" meant "a large formation of enemy aircraft."

blitz bomber — late in WWII, a term applied to the new German jet bombers, such as the Arado 234. — Monty Berger.

block time — elapsed time from the moment the aircraft begins to move off the ramp, until the time it comes to rest again at destination. Derived from the removal and replacement of blocks (chocks) for wheels. See **chock-to-chock**.

blockbuster — in WWII, a 2000 pound bomb. See also **cookie, tallboy, Grand Slam, doodlebug.** —from Doug Sample CD, Yorkshire Air Museum, Canadian Branch.

blood box — ambulance. See **blood wagon**.

blood wagon — WWII for ambulance. Also known as the **meat wagon**. Current slang has it as **blood box** because of the boxlike shape of many modern ambulances.

blower — telephone. WWII. Also, more recent jargon for supercharger.

blown flaps — fixed-hinge, slotted-vane flaps deflected into the engine exhaust to produce reaction lift augmentation. The system also incorporates full-span leading-edge slats and spoilers, all to give added lift at low flying speeds, such as during short-field landing approaches or air-drops. Also called **powered flaps**.

blow show — among "stewardesses," "air sickness," according to John Davis, *Buzzwords*.

blow-start — starting a jet engine by blowing exhaust from another jet directly into the intake. —from Chuck Yeager's autobiography.

Blue Angels — the U. S. Navy aerobatic team. See **Thunderbirds, Red Arrows, Snowbirds.**

blue ice — "apparently frozen toilet waste" from airplanes aloft. A block of "what is known in the aircraft industry as 'blue ice' crashed through the bedroom ceiling of a home in a Toronto suburb on Canadian Thanksgiving Day, 1994, according to a story in the *Globe and Mail*. "Airplane toilets are drained on the ground, but a leaky drain valve can allow a cluster of ice to build up during high-altitude flights. At some point, often during a landing approach as the aircraft enters warmer temperatures, the ice breaks off." It only happens about once a year, but nobody keeps statistics on the exact frequency. —"Aircraft's frozen dropping terrifies Toronto householder," *Globe & Mail* (October 13, 1994).

blue letter — letter of complaint about a flight attendant. — Paul Dickson, *Slang!*

blue-on-blue — "clear, blue skies over a clear, blue sea; calm weather." —Murray, "....Navy Fighter Pilots". See opposites: **muddy, clag.** However, John and Adele Algeo, "Among the New Words: Gulf War Glossary," *American Speech* (Winter 1991) define it as "friendly fire (allusion to the Civil War)"; and Alec McRitchie, public relations for Shorts aircraft manufacturer, Belfast, confirms this sense—"friendly fire taking out an aircraft."

blue room — toilet on an aircraft. — Paul Dickson, *Slang!* See also **three-holer.**

Blue Shirt — U.S. Navy term for aircraft carrier crewmember in charge of the mechanical components of an aircraft, so named because of the color of his shirt. See **Yellow Shirt, Brown Shirt, Red Shirt.**

B'nai — Hebrew for twinned, an Israeli expression for flying in formation.

boards — speed brakes.

bods — RAF WWII slang for "bodies", service personnel.

Boeing — "At one period in the 1950s the Air Force was operating so many Boeing types, B-29, B-47, B-50 and B-52, that Pentagon wags would facetiously explain that the 'B' actually stood for Boeing rather than bomber or bombardment." —Carl Cleveland, *Boeing Trivia*.

Boeing Brewery — jocular name arising out of an event in Boeing public relations: "Vogel was in charge of a visiting press group...from Australia....Vogel arranged a trip to Mt. Rainier (**Mt. Boeing**). Aware of the Australian propensity for keeping the whistle wet, Vogel had stocked the two station wagons with a generous supply of beer, kept cool with a substantial supply of ice.

'We weren't more than 10 miles south of Renton ... when our guests noted the beer....We pulled over to the side of the road and started handing out the stubbies. The first Aussie to taste his beer sputtered and exclaimed in disgust, "This stuff is cold!" The others joined in voicing a decided objection to American cold beer.

'They solved the problem by lining up the bottles on the hoods of the station wagons in the sun and waited for a proper "warming." Many goggle-eyed passing motorists slowed to get another look at the "Boeing Brewery" at roadside, until the Aussies decided the beer was at a drinkable temperature.'" —Carl M. Cleveland, *Boeing Trivia*.

boffins — RAF WWII slang for **back-room boys** and scientists.

bogey — 1. "enemy plane" — Murray, "....Navy Fighter Pilots". Also "any unidentified flying aircraft" or, for that matter, any mysterious blip on a radar screen. But in HW's *What's the Gen?*, a 1942 pamphlet of RAF cartoons and slang terms, "bogey" is glossed as "friendly aircraft." 2. wheel assembly.

These two different meanings come from two different old English/Scots words. The Oxford English Dictionary identifies "bogy" as a wheeled conveyance or 'truck' from far back; it dates from the 19th century use of "bogey" for "evil one, devil, goblin, person much dreaded" and from 15th century Scots the related "bogle," meaning "phantom, thing insubstantial."

bogey country — "enemy territory; airspace where enemy planes are thought to be." —Murray, "....Navy Fighter Pilots".

bogey dope — "bearing, range, altitude, and number of unidentified aircraft." — Murray, "....Navy Fighter Pilots".

boing — pidgin word for "jet airplane," derived from "Boeing."

bolter — an aircraft which fails to catch an arresting wire while landing on a carrier, therefore running off the end of the landing deck. Pilots routinely give full throttle momentarily after touchdown and wait for the deceleration that indicates they have caught the wire; if they haven't, they have full power ready to abort, take off, and go around again. —Coonts, *Flight of the Intruder.*

bomb aimer — British term for **bombardier**.

bombardier — American term for the bomber crew member responsible for guiding the aircraft over the target and actually releasing the bombs. Known in Britain as the **bomb aimer**.

Bomber Harris — Air Marshal Sir Arthur Harris, chief of RAF Bomber Command during WWII. According to one source, aircrews nicknamed him 'Butch' because of his bulldog tenacity. Critics called him "Butcher" because of his ordering bombing of civilian targets. —Spencer and Harris, *Reap the Whirlwind.*

bomber's little friends — what WWII bomber crews called their fighter escorts, mainly applied to P-51s, P-38s, and P-47s. Also called **little friends**.

bomber's moon — one that wanes at the time of arrival in **bogey country**. WWII.

bomber's night — "Moonless night, such as 17 January 1991 over Iraq, when the air campaign began" —From John and Adele Algeo, "Among the New Words: Gulf War Glossary," *American Speech* (Winter 1991).

bombing up — taking on a bomb load in preparation for an attack.

bonk — crash a hang glider. **Whack.**

boom — the refueling probe mechanism at the rear of a tanker aircraft (e.g. USAF KC-135). The boom operator, they say, is "the only enlisted man in the air force who gets driven to work by two officers."

There is a story, unverified by us, that a USAF KC-135 tanker once towed an engine-out F-4 fighter aircraft most of the way across the Atlantic, with the boom, separating and allowing it to glide to a dead stick landing at Loring AFB, Maine.

boost override — a safety stop that must be overcome to open the throttle full-wide (British preWWII).

boot chute — among "stewardesses," the emergency exit, according to John Davis, *Buzzwords*. The term is likely derived from the necessity for flight attendants, in an emergency, to "boot" the passengers out of the aircraft.

boozer — radar warning device against fighters, fitted to bombers in WWII. —Doug Sample CD, Canadian Branch, Yorkshire Air Museum.

boring holes — flying aimlessly or without destination. More fully "boring holes in the sky." —from Paul Turk. Also **honking about**, **swanning**. See also **punching holes**.

bottom rudder stall — characteristic of the Vultee BT-13 Vibrator. According to Bill Kennedy Jr. in *Air Classics Magazine* (May 1980) "...a spin entry out of a turn. Every student had this manoeuvre demonstrated to him by his instructor and had discussed how it could happen. The classic example..... insufficient bank causing the end of the runway to slide by the nose. The student would then start feeding rudder into the turn to get the nose around and as the nose started to go below the horizon he would feed in **back stick**. Bingo! Down would go the inside wing and around she'd go. Since this occurred on the final turn, there was no room to recover unless you snapped it all the way through and recovered when you were once more right side up."

bought it — was killed or shot down. From "bought the farm." WWII. The area beyond the runway is often an open field and is known to fliers as the **farm**. See **buy the farm**, **auger in**, **go for a Burton**.

Bought on American Credit — irreverent nickname for British Overseas Airline Corporation, before privatization. Later BOAC was explained as **Better on a Camel**, **Bring Over American Cash** and **Bend Over Again Catherine**.

bounce — to jump or be jumped, i.e. to attack without warning, or be so attacked. "To attack an enemy airplane from the air, usually from a higher altitude, taking the enemy unaware." —Woodford Agee Heflin, ed. *The United States Air Force Dictionary*. 2. British term for a **touch-and-go** landing.

bouncing bomb — developed in 1943 by Sir Barnes Wallis for attacks on German dams in the industrial region of the Ruhr and Eber valleys, this device was held in a set of brackets which would cause it to ricochet off solid objects (like land or dam) and tumble into water, where it would explode at 50 feet of depth, from pressure. The bomb would skip about three times off water, and had to be placed precisely about 200-400 feet in front of the dam; its aim was to create water pressure to break the dam. And it did. The loss of hydroelectric power crippled the German war effort. (And the floods overran at least one POW camp, killing hundreds). See **Talking Bomb** and Paul Brickhill's *The Dam Busters*.

Le **Bourget** — common name for the Paris International Air Show, held every two years (1991, 1993, etc) and considered the premiere of these international exhibitions and trade fairs: the 1995 version was the 41st. So known because it is held at the business/private airport north of Paris located on the site, at the field, Le Bourget, where Lindbergh landed after the first transatlantic solo flight. As early as 1910, it was envisioned as a site for an aerodrome, and was the main Paris airport for many years.

bowler hatted — RAF WWII term meaning **washed out**, failed flying school, ordered to **"cease flying"**, sent back to **civvy street**.

bowser — tank truck, for refueling aircraft, in England during WWII. Often in the form "petrol bowser." —Monty Berger.

box barrage — an enveloping anti-aircraft barrage around an objective or target.

boxkite — common name for early aircraft design which was literally based on box kite principles, typical of early Wright, Farman and Voisin designs. Also applied generically to any early aircraft. See **bird cage**.

box office — jocular name for the **cockpit**, some said, after the introduction of women as Air Canada pilots. See also **empty kitchen**.

Bradshawing — In early British aviation, navigating by using the British railway timetable, published by Bradshaw, which all pilots carried, and following the railway or using it and visible trains to determine position. See **iron beam**, **iron compass**, **hit an open switch**.

brain bag — the bag in which a pilot carries his charts, computer, plotter, checklist, orders, instructions, etc.

brain bucket — crash helmet.

brats — RAF Volunteer Reserve or Regular airmen who had been through RAF Apprentice School for technical trades. They began training when aged between 15 1/2 and 17, having committed to remain in the RAF until age 30. —Deighton, *Fighter*.

Bravo Delta — "Broke Dick," from the phonic code for a piece of hardware that doesn't work. —Hans Halberstadt, *Airborne: Assault from the Sky*. See **Tango Uniform, Uniform Sierra**.

break — "change course (said of an airplane)." —Murray, "....Navy Fighter Pilots". Usually a violent, turning evasive manoeuvre.

2. A term used in British naval aviation to indicate a point near the runway or carrier deck where a flight of aircraft "breaks" into sections or individual aircraft in preparation for landing.

breakaway — USAF phrase to describe an emergency manoeuvre meant to quickly separate aircraft vertically during an in-flight refuelling.

break-away — the initial movement of an aircraft from the static or parked position. See **push-back**.

break-away height — British term describing the minimum height below which a landing approach cannot be safely continued. Also called **break-off height**.

break-in — said of new or overhauled engines, usually the first 50 hours of operation, when the **mill** needs special care and handling. See **run-in**.

breakout — to break through the clouds, usually when descending to land. See **on top**. .

briefing — information session. See also **debriefing**. —Doug Sample CD, Canadian Branch, Yorkshire Air Museum.

broke a spoke — "engine failure," among "stewardesses," according to John Davis, *Buzzwords*.

brolly — a British slang term for a parachute, taken from "umbrella" in British slang. See **chute, silk canopy**.

Brooklands — eventually applied to identify a kind of British racing car of the 1930s and 40s, the term comes from the name of an old airfield and racetrack with banked turns and long straightaways, approximately oval in shape, south of London. The race track was built near the turn of the century, but its singularity emerged during the 1930s because of the experimentation with aircraft engine technology. The track enclosed a flying field. People like Harry Hawker started their factories here. Vickers factory was here during the war. Near the Riley aircraft engine factory, it was the site of races between Napier and Railton race cars, the former powered by the Rolls-Royce Lion V-12 and H-24 engines, and the Rolls-Royce Merlins, with 24-cylinder supercharged engines also used in the Spitfire. Some were V-12s and developed 1100 horsepower. The Indy 500 of England. See also **Allison V-12.**

Brown Shirt — U.S. Navy term for an aircraft carrier crewmember in charge of the launching catapult, so named because of the color of his shirt. See **Blue Shirt, Red Shirt, Yellow Shirt.**

brown shoes — shorthand for "a member of the U. S. Army Air Corps," from their distinctive footwear, and because they had no special flight uniform. — Jess Finney, former USAAC, San Diego.

2. "British slang for naval aviators." — E. B. Ocran, *Dictionary of Air Transport and Traffic Control.*

brownout — loss of orientation and/or visibility due to helicopter downwash creating localised dust/sandstorm. See **blackout, grayout, whiteout, redout.**

buck — to fly against the weather, as in "buck the **muck.**"

bucket — USAF slang for a jet engine turbine, Korean era. —Glen Infield, "The Day Ted Williams Almost Got It," in *Fighting Eagles,* ed. Phil Hirsch.

bucket shop — an archaic term from the Stock Exchange used to describe unscrupulous travel agents who sell tickets below legal or recognized prices.

Buckingham — a type of machine gun ammunition used by British flyers in WWI. See also **de Wilde.**

buddy-buddy pack — flight refuelling equipment carried in a streamlined container on a weapons pylon. See **buddy store.**

buddy plane — "aircraft that uses its laser beam to control a bomb dropped by another plane" — From John and Adele Algeo, "Among the New Words: Gulf War Glossary," *American Speech* (Winter 1991).

buddy store — a detachable fuel tank which can be carried by one aircraft for the purpose of refuelling another aircraft. See **buddy-buddy pack.**

buff driver — pilot of a **BUFF** (Big Ugly Fat Fucker or Boeing B-52 — see "Plane Names").

buffet — violent shaking produced in rough air or at very high speeds. In the latter case, it is the "speed of sound" factor, or what was called the point of maximum **compressibility,** in terms of air resistance. 2. British term for the aircraft galley.

buffet boundary — the point at which there's enough thrust but not enough wing.

bugeye — 1. applied to aircraft with bulging cockpit side, top, or front windows, such as the Grumman OV-A Mohawk.

2. on the Bellanca Cantabria, the air intake on the cowling.

bug eyes — side-by-side plexiglas canopies atop certain aircraft to separately enclose both members of the flight crew, as on the first U.S. jet bomber, the Douglas XB-43. See **blisters**.

bug smashers — 1. small private aircraft, especially in congested areas. — Paul Dickson, *Slang!* See also **Indians**. 2. a propeller

build-up — a mass of cloud, towering cumulus, thunderclouds, stormclouds, usually to be avoided.

bull nose — See **bunny nose**.

Bullseyes — night affiliation exercise (bomber training flights over British cities to learn to dodge searchlights, night fighters, and anti-aircraft guns. WWII.

bumf — WWII term for leaflets dropped from aircraft. Also used to refer to official correspondence, propaganda, news releases.

bumper — among "stewardesses," a clumsy passenger, according to John Davis, *Buzzwords*. See also **Etna, poker, earache, claw, faucet, sweater, tenpins**.

bump load — "to not load something on a flight and instead load it on the first available flight." —Mary Manni, "Say What? A guide to airline slang," Air Canada's *Horizons Magazine*, April 27, 1994

bumps — See **air pocket**.

bunch of bananas — Luftwaffe term for the usually-flown RAF **vic** or **vee** formation, which they considered obsolete and highly ineffective. —Deighton, *Fighter*.

bungee muscle — bungee cords attached to the control column to help steady a badly out-of-trim aircraft, in the good old days.

do a **bunk** — leave without permission, go AWOL. WWII. —Doug Sample CD, Canadian Branch, Yorkshire Air Museum.

bunny nose — short rounded nose often seen before nose radomes became standard, such as on early Piper Apache, Cessna Bobcat or Beech Expeditor, as opposed to long, pointy nose. Sometimes called a "bull nose."

bunt — British for a sudden dive. See **pitch-out**.

bunting — flying low over hills, trees, under power lines, etc., using ground features as cover. A Gulf War term. —U. S. News & World

Report, *Triumph without Victory*, 1992.

burn-off — the amount of fuel consumed by an aircraft. Also called fuel burn.

burst speed — a noun, used to describe what you get when you want maximum speed fast, using **RATO, JATO, firewalling it, afterburner, panic boost, through the gates**. See also **dash modes, dash speed**.

Burton, went for a — was shot down, killed. (Burton's was a brand of ale sold in Britain. British and Canadian usage) See also "got the chop", **pranged** —from Doug Sample CD, Yorkshire Air Museum, Canadian Branch. See **auger in, buy the farm**. Another story traces the phrase to Blackpool, England where RAF riggers, mechanics and gunners were trained during WWII. The base had no Guard Room or **glasshouse**, so offenders were held in a room above a tailor shop, "Messers Burtons - The Fifty Bob Tailors". To those stationed at Blackpool 'gone for a Burton' meant thrown in the clink, but was later expanded to apply to a missing squadron mate. —Hooper, *Pilot Officer Prune's Picture Parade* (1942).

bush plane, bush aircraft — "The term 'bush aircraft' like the term 'bush flying' had been used since the very early days of bush flying in Canada but it is not really capable of definition. Obviously a bush aircraft is one that has been used for bush flying, but some types were used which were not really suited to it.... A bush aircraft ... could carry a payload for a reasonable distance together with enough equipment to take care of itself and its crew under emergency conditions during winter and summer. It should be rugged, easy to maintain and capable of landing and taking off in a small area. Speed is not the prime consideration." — Molson, *Pioneering in Canadian Air Transport*.

bust — exceed limits of air traffic control clearance. One is said to have "busted clearance."

busted forecast — inaccurate meteorological report or sudden, unforecasted weather.

buster — command or condition requiring the opening of throttles fully, **through the gates** to full emergency power, such as in a **scramble,** or to escape superior enemy fire.

butterfly — British slang for helicopter. See also **chopper, whirlybird**.

butterfly tail — Vee-shaped tail utilising a **ruddervator** control surface which combines functions of the rudder and elevator, as on the Beech Bonanza. See **V-tail**.

buttering — in the work on the Lockheed F-117A stealth fighter, "the airplane's special need to have absolutely smooth surfaces in order to maintain maximum stealthiness caused unusual stress for ground crews. After each flight the radar-absorbing materials had to be removed to gain entry to doors and service panels, then had to be meticulously replaced in time for the next mission...The process was called 'buttering,' using a special radar-absorbing putty we developed to coat uneven surfaces." — Rich, *Skunk Works*.

button — the **threshold**, the marked beginning of the runway. "The landing gear would be about ten feet over the runway when he crossed the button." — Beaudoin, *Walking on Air*.

buy the farm/ ranch — to die in an air accident, especially a pilot. See **auger in**, **go for a Burton**.

buzz — a manoeuvre, usually illegal, whereby the airplane swoops low and fast over a selected location, such as the control tower, the commander's office, or the girl friend's house. **Beat up**.

buzz bomb — British term for the German V-1 rocket, the **Doodlebug**.

buzz job — (verb) "to fly low, 'cut the grass.' Final manoeuvre practiced by many pilots no longer with us." — Stevens, *"There I was."* See **balls, low flying**.

buzz numbers — big numbers required to be on the side of airplanes, dating from post WWII, to make it easier to identify and prosecute pilots who liked to **buzz.**

CAAC — officially Civil Aviation Administration of China, the flights it operates have earned the insulting variant "China Airlines Always Cancelled." — Jan Wong, "Plane Rude: Service with a snarl," Toronto *Globe and Mail*, December 4, 1993.

cabotage — "carriage of traffic between points within one and the same territory (generally reserved for the national airline). —Swissair's *ABCs of the Airline Industry*.

CAG — USN Commander, Air Group , redesignated "air wing commander" in 1962, but still referred to as CAG. See **double nut.**

caged bird — a grounded aviator or one on desk duty. See **flying a desk**.

can — RAF WWII name for the container holding incendiary bombs. When dropped over a target, a can would normally break open, scattering the incendiaries for maximum effect on the ground. See **Molotov breadbasket**.

Can Air — One name proposed for a merged Air Canada/Canadian Airlines conglomeration. (Gord Sinclair, Montreal radio station CJAD) Punsters quickly converted "Can Air" to "Tin Can Air." See also **Mapleflot**.

canards — French for "ducks," small, independent aerodynamic surfaces used for aircraft control. Well forward, ahead of the CG, (makes the plane look like a duck in flight). First applied to biplanes. "Tailplane at the front, for greater stability." On the Mirage jet, known in French as **moustaches**.

canoe — container attached beneath fuselage to carry photo equipment and infrared sensors, such as on the North American-Rockwell A-5 Vigilante.

CAP — Combat Air Patrol. 2. Civil Air Patrol. See **BARCAP, SARCAP**.

capped him — "flew Combat Air Patrol over him...." From remarks made by General Horner, U.S.A.F. commander of allied air forces in the Gulf war, describing what a flyer did to protect one of his downed comrades......(at the 1992 Aviation/Space Writers Association conference in Montréal). Confirmed by John and Adele Algeo, "Among the New Words: Gulf War Glossary," *American Speech* (Winter 1991), who, however, list it as having "UK and Canadian source." See **top cover, BARCAP, SARCAP**.

captain — term for chief pilot, especially of a commercial airliner. Can be identified by the four stripes on the sleeve of his uniform. For some of the implications of this choice, see **ship**.

captains of the clouds — poetic description for aviators.

carpet bombing — method of bombing in which target area is saturated with bombs in hopes some will hit actual targets.

carrier landing — "Heaven on deck starts by connecting with number 3 wire." — Foster, *Sea Wings*. Also **home and dry, land on, tailhook, bolter.**

casevac — casualty evacuation, term used in Korea. Most cases were evacuated by air, Korea seeing the first combat use of helicopters. See **medevac**.

Castle Dismal — term used to refer to No. 6 (RCAF) Group Bomber Command Headquarters at Allerton Park, England (a 75-room Victorian mansion) —Doug Sample CD, Canadian Branch, Yorkshire Air Museum.

cat and mouse — the two signals comprising the Oboe bomber navigation signal of RAF WWII. The "cat" signal guides the aircraft to fly a circle of constant radius from it, while the "mouse" signal intersects the "cat" signal over the target. When the pilot reaches the intersection of the two signals, he is in position to drop bombs.

Caterpillar Club — made up of people who have bailed out to save their lives. If the caterpillar in the pin has red eyes, the plane was burning. The name "caterpillar" is elucidated by knowing another term for such survivors, **silk bloomer.** Parachutes were originally made of silk. —Berrey and Van den Bark, *The American Thesaurus of Slang.* Doug Sample recalls that the "small gold caterpillar was issued by the Irving Parachute Company," a great use of a slang term for public relations. See **Flying Boot Club, Walker's Club.**

cat shot — "catapult takeoff from an aircraft carrier" — From John and Adele Algeo, "Among the New Words: Gulf War Glossary," *American Speech* (Winter 1991). See **cold cat shot, squirted off.**

CAVU — "Ceiling and Visibility Unlimited."

CCIT — Controlled Crash Into Terrain. —from Paul Seidenman, aviation writer, San Francisco. Also **CFIT**, Controlled Flight Into Terrain, a term applied to aircraft which appear to have flown straight into the ground.

"cease flying" — official term used by instructors to indicate student was not qualified to finish a flight training course. RCAF WWII. — Williams, *The Plan.* **Washed out, bowler hatted.**

ceiling — before its now universal use to designate the altitude at which cloud overcast reduces visibility to zero, the "bottom of the clouds," it was a term for "altitude limit," with the "spec. 'floor' — the rarefied stratum of upper air in which an airplane can barely sustain flight.'" —Berrey and Van den Bark, *The American Thesaurus of Slang.*

ceiling chicken — "the person who goes up the baggage belt to free something that is jammed in it." —from Mary Manni, "Say What? A guide to airline slang," in Air Canada's *Horizons Magazine*, April 27, 1994.

center-of-gravity (CG) — the balance point, affected by loading, attitude. That which must be found and maintained in order to attain and maintain flight. "You distribute the loads (people, cargo, fuel, munitions, whatever) to ensure that the resultant **CG** is close to a point below the point where the wing lift acts. The horizontal stabilizer (US term) or tailplane (Brit.)

makes up for any out-of-balance condition, but it can only do so much, so the CG range for any aircraft is critical; too far aft, and you'll get off the ground (for a very short time!); too far forward, and you'll become a high-speed ground vehicle unsuited to the road system." — anon. Canadair technical writer informant.

centurion — British naval term for a pilot who has made at least 100 carrier landings.

a **CEO-is-aboard-landing** — see **kiss landing, greasing it in.** — from *Flug Revue*, #2, June 1992.

CFIT — Controlled Flight Into Terrain. Also **CCIT**, Controlled Crash Into Terrain.

CG — center of gravity. "Canadair corrected a CG problem by increasing the area of the elevator."

chaff — thin strips of metal foil dumped from aircraft to fool radar. See **window**.

Chance Light — powerful light at the end of the runway to assist in take-off and landing (WWII military aviation). Made by Chance Bros. —Doug Sample CD, Canadian Branch, Yorkshire Air Museum.

chandelier — RAF WWII term for a large formation of aircraft at great height, usually providing **top cover** for formations below.

chandelle — an abrupt climbing turn in which extra altitude is gained by the momentum of the plane which approaches stalling angle during the change of direction. — Williams, *Casey Jones Cyclopedia of Aviation Terms*.

char — WWII term for tea. —Doug Sample CD, Canadian Branch, Yorkshire Air Museum.

Charlies — USN slang for clearances to land. —Coonts, *Flight of the Intruder*.

chart — a map, more properly an aeronautical chart showing a plethora of information of use in pilotage.

charter airlines — companies that run excursion flights, mainly, as opposed to airlines with regularly scheduled flights. Their ease and frequency of creation and liquidation has led to the question, "What's the difference between an airline and an airplane? An airplane can stay off the ground longer." —from "View from the Cheap Seats," Montréal *Gazette*, August 1, 1992. Also called a **non-sked**.

chase plane — an aircraft flown in close proximity to another to observe its performance during testing or training.

chasing the airspeed — attempting to maintain straight and level flight by watching the airspeed indicator, usually resulting in **porpoising**.

chasing the needle — sloppy flying, or flying in turbulence when it's hard to keep the "needles" where you want them.

checking the tailwheel — expression sometimes used by flightcrew to explain their frequent lavatory trips to the passengers. Because of constant travel, many aircrews, especially during the DC-3 era, when drinking water was often of questionable quality, suffered what has been variously described as Memphis Murmur, Chicago Challenge, Houston Hopscotch, Denver Dance, Texas Two-Step and, of course, Montezuma's Revenge. DC-3 facilities were wedged 'way at the back of the cabin. Polite modern term; gastrointestinal distress. See **taildragger**.

checklist — there's one for pre-start, startup, taxi, takeoff, climb, cruise, descend, land, shutdown, emergency and a host of others as airplanes and their loads become more sophisticated. One informal list: "Kick the tyres, light the fires." See **PUFF check, GUMP check, preflight, patter**.

check ride — flight on which a pilot or an aircraft are being put through a series of tests. In the case of the aircraft, it is usually known as a "test flight."

check six — "Pilots' catchphrase for 'look directly behind you (at six o'clock) to be sure no enemy is following." —From John and Adele Algeo, "Among the New Words: Gulf War Glossary," *American Speech* (Winter 1991).

cherry jumpers — paratroopers in training, before their sixth jump. See also **snuffies, prop blast party**.

cherubs — measure of altitude in hundreds of feet are known as **cherubs**, a slang term derived from **angels**.

chicks — "friendly fighter aircraft." —from a Canadian Forces Aircom Regulation Glossary.

chiefy — flight sergeant (WWII). —Doug Sample CD, Canadian Branch, Yorkshire Air Museum.

Chinese landing — Wan Wing Lo ("one wing low"), a sloppy landing. Also descibed by Henry L. Williams, *Casey Jones Cyclopedia of Aviation Terms*, as being a downwind landing.

CHIRP — Confidential Human Factors Incident Reporting Programme - an agency to which an aircrew member can confidentially report human errors or inadequacies noticed in oneself or fellow crew members, or in procedures or equipment, for follow-up, evaluation and improvement without fear of reprisal.

chisel charter — "An illegal bush plane charter, usually done by a private pilot without a charter license. The costs are much lower than those of a legitimate charter operator." — Parkin, *WetCoast Words*. See also **bandit**.

chock-to-chock — flight time, i.e. from the time the wheel chocks are removed to the time they are put in place at destination. See **block time**.

chop — "light, regular turbulence, usually not enough to spill a coffee." —Mary Manni, "Say What? A guide to airline slang," Air Canada's *Horizons Magazine*, April 27, 1994. Can be distinguished as "light," "moderate," or "severe." "Center, **Speedbird** 1-1. We're getting a little chop at 3-5. Can we try higher?" See **airpocket**, **ride**.

chop, got the — British and Canadian WWII way to say "was shot down." See also **Burton, went for a** and **gone in the oggin**.

chopper — slang for helicopter, **copter, egg beater, sling wing, whirlybird**.

chuff — see **chaff, window**.

cigarette roll — a particularly vicious malfunction in parachuting, which occurs when the lines are twisted together by the 'chute pack's having rotated instead of deploying directly; the lines then are fused by the friction into one inseparable line, preventing the parachute's opening and practically guaranteeing the parachutist's being **first one on the ground**. British (RAF) term is **Roman candle**.

circle jerk — see **cluster fuck**.

circle trip — travel agent parlance for round trip.

circuit — the traffic pattern, usually around smaller airports. Aircraft wishing to land enter the pattern, fly **downwind**, then turn on to the **base leg** at right angles to the runway, then on to **final** where they line up with the strip and land. Can be flown either right-hand or left-hand. See **circuits and bumps, bounce, final, touch and go**.

circuit riders — flying doctors who visit small towns and remote communities on a regular or random basis by air. See **fork-tailed doctor killer, doctor in a Bonanza**.

circuits and bumps — WWII term for practice landings and take-offs. —Doug Sample CD, Canadian Branch, Yorkshire Air Museum.

See **bounce, rollers** and **touch-and-go.**

circus — a fighter-escorted daylight bomber sortie by the RAF in WWII. See also **fishpond.** —Doug Sample CD, Canadian Branch, Yorkshire Air Museum. "The circus was intended as a feint to draw off German interceptors. As the main striking force flew to the primary target, a smaller group of bombers with fighter escort attacked a diversionary target." — Robert S. Johnson, *Thunderbolt.* Flying Circus was the name given to Baron Von Richthofen's Jasta II squadron during WWI, so named because of their practice of painting their **Albatri** in bright colors and outrageous designs. The term is also generally applied to any mass or formation of aircraft.

civil aircraft — privately or corporately owned aircraft, i.e. not military, but including commercial.

civvy street — civilian life. WWII, all branches of service. Ward-Jackson's *It's a Piece of Cake or R.A.F. Slang Made Easy* identifies the origin of this word and its various combinations as being during the "Great War." "Wearing civvies" meant out-of-uniform. See **bowler hatted.**

clag — "bush pilot's term for calm coastal weather of low cloud, fog, and drizzle. According to Bob Dalgleish, former flyer, 'It was easy to get lost in clag because one had to fly low over the water, so it was hard to keep a reference point as the mists blotted out distances and distorted landforms. Occasionally one had to land and ask a fisherman, who was hopefully slightly less at sea than you, where exactly you both were.'" — Parkin, *Wet-Coast Words*. See **duff.**

clanks — nervousness. "I'm not brave, I've got the clanks." —Bob Stevens, *"There I was."* See **twitch.**

claw — "first-time flyer," among "stewardesses," according to John Davis, *Buzzwords.* See also **faucet, sweaer.**

clean — gear, flaps, spoilers, air brakes, etc., in up position for maximum streamlining. See **dirty.**

"clear!" — the standard warning shouted before starting engines. In German it is "Frei!"

clear air turbulence — "occasionally, on a clear day, unusual wind conditions will cause a bumpy flight. This unusual occurrence is called 'clear air turbulence.'" —"Airline Taalk," American Airlines Field Training/Ground Services materials, March 1993.

clear deck — British naval aviation term indicating the deck is ready to receive landing aircraft. Opposite of **foul deck**. See also **ready deck, green deck**.

cleared short — given ATC **clearance** only part way to destination, expecting further clearance enroute.

clearance — ATC permission to take-off, land, enter or leave airspace, fly at a given altitude, speed, heading, etc.

clearance roll — illegal showoff manoeuvre. During descent, pilot rolls inverted to check traffic below. —from Chuck Yeager's autobiography. See **victory roll**.

clear perspecs — "of the turret that you cleaned and recleaned this afternoon." Clear vision. "Clear Perspex" is a brand name of a toughened glass. "What did you use to clean the perspex?" "The No. 7 rag wrench, of course!" —Jess Finney, former USAAC, San Diego. In the U. S. the term is **plexiglas**.

climb and glide — a coordination exercise including glides, gliding and climbing turns in two directions. — Williams, *Casey Jones Cyclopedia of Aviation Terms*.

clock — the altimeter, as in "There I was, upside down, nothing on the clock but the maker's name.....!" — Stevens, *"There I was."* Also can be applied to other instruments: "But the sudden, gleeful dive had given him a bigger increase in speed than he'd bargained for. Nearly 140 on the clock - fifty-five m.p.h. above the normal landing speed. He was going to put her down all the same!" — Forrester, *Fly For Your Life*. The expression can and is used generically for virtually any instrument. See **on the clocks**.

clooge — "especially among missile men, any improvised or makeshift repair, using old tin cans, chewing gum, etc., to simulate first class work in Tech Order compliance. Usual results will not pass 'eyeball.'" — Stevens, *"There I was."* This word, which also turns up in car and motorcycle slang, comes from a German word for "clever," "klug," which is often used ironically. See Poteet and Poteet, *Car & Motorcycle Slang*.

closed — on the airport information board it means the doors of the aircraft are closed and the passenger has missed the flight. On **NOTAMS** it means the airport or runway is closed due to weather or obstructions. Referring to **flight plans**, it means the aircraft has arrived and the **ATC** has been notified.

close-air-support — Ground support aircraft. "plane such as the

Warthog" — From John and Adele Algeo, "Among the New Words: Gulf War Glossary," *American Speech* (Winter 1991).

closed box formation — bomber formation adopted by American bombers in WWII calling for five aircraft to fly in tight formation, thereby offering maximum protection for each other against enemy fighter attacks. While it offered optimum positioning for protective machine gun fire, the formation had its drawbacks in that the tightly grouped airplanes presented **flak** gunners with a better target and the likelihood of collision was increased.

closet hopper — among "stewardesses," a male steward, according to John Davis, *Buzzwords*.

Clouds — nickname for the meteorological officer, USAF. —Coonts, *Flight of the Intruder*.

coach-seat clot — "Blood clot caused by sitting still for a long period of time, as in an airplane seat." —"Among the New Words, *American Speech* (Summer, 1995). See also **economy class syndrome**.

coast-in point — the point at which an aircraft will cross the coast, going from over water to over land, **feet wet** to **feet dry**.

—Coonts, *Flight of the Intruder*. See also **coast-out point**.

coast-out point — point at which an aircraft will cross the coast going from over land to over sea, **feet dry** to **feet wet**. —Coonts, *Flight of the Intruder*.

cobbed it — push it to the limit of available power. Taken from motorcycle slang, where the twist-grip accelerator was known as the "corncob" (see Poteet and Poteet's *Car and Motorcycle Slang*. See **firewalled it, through the gates**.

cobra — aerobatic manoeuvre, best performed by the MiG, in which the aircraft does a **tailstand**, then comes out of it under power and brings the nose down in a striking motion similar to the strike of the cobra snake.

cockpit — the "flight deck," the compartment at the front of the aircraft housing pilots and controls. Also known as the **office, playpen, pulpit** and, jocularly, as the **boxoffice**, and "the area where chicken pilots are kept."

cockpit check — procedure during which pilot knowledge of procedure, equipment, and layout are verified before takeoff.

cockpit management skills — see **stick and rudder**.

cockpit on a stick — nickname applied to a specially built test vehicle during Boeing's development of the Boeing 777. Its official name was the LAGOS, for Large Airplane Ground Operation Simulator. It was a cockpit mounted over nose wheels and attached by a long beam to the other running gear in the configuration designed for the aircraft, so that they might test its manoeuvreability on taxiways and around airport ramps.

cockpit queen — "a flight attendant who fraternizes with the pilots," according to "L.A.Speak" in the Los Angeles *Times* magazine, March 20, 1994.

coffee grinder — old term for the old radio systems that were tuned with a crank and flexible shaft.

coffin corner — 1. slang expression for very high altitudes, unsupportive of wing lift, a real danger to certain aircraft (e.g. old **T-tail** Learjets). At these altitudes the thin air, working less effectively with the **critical wing**, may conspire to put the aircraft into a **flat spin**.

2. In engineers' slang, "combinations of speed, altitude, and temperature...at the extremes of the envelope...that you stay away from if you possibly can." —Sabbagh.

C O D, cod — Carrier Onboard Delivery, generic term applied to carrier-based transport aircraft. US Navy. Paul Dickson calls "cod" a "plane that delivers mail and other supplies to an aircraft carrier."

co-in — counter insurgency. Term applied to ground support aircraft during the 1960s, such as the Cessna A-37 jet trainer cum attack vehicle.

Coke bottle design — see **area rule, Marilyn Monroe design.** The Lear 45 is the first Learjet not to feature this fuselage shape.

cold cat shot — a catapult shot from a carrier's bow, in which "the catapult fails, usually dribbling the plane off the bow." — Searls, *The Crowded Sky.*

Cold-Nosed Order of Arctic Pilots — a confraternity founded by legendary bush pilot Matt Berry, comprising pilots who have flown north of 70 degrees north latitude. —Munro, *The Sky's No Limit.*

cold-soaked-fuel — fuel in wing tanks warms slowly after landing. While on the ramp, the cold fuel may cause airborne water vapor to form ice on the wings, which is usually shed on take off, but sometimes requires de-icing.

"Those of you who fly the [Douglas DC-9] S80 regularly know about de-icing, even when the weather is balmy. The S80's fuel is stored in those skinny little wings, and at altitude, the fuel gets pretty chilly. On the ground, moisture in the air periodically forms ice on the wings (a) causing de-icing procedures to be performed as other airplanes taxi by, crews laughing hysterically (b) the Captain has the First Officer check the wing surface for ice by licking it, if he sticks it is a "**no-go**", (c) passengers are startled to see Nancy Kerrigan doing a triple Lutz on the wing, or (d) all of the above. The answer of course is all of the above." — Don Loughran in *News Aloft*.

collect a gong — be awarded a decoration (medal). WWII.

colours — see **livery, white-tail.**

COMAT — "Company material," in code. The "Company Pouch."

combi — commercial aircraft which routinely carries both passengers and freight, derived from "combination."

commander's moon - U.S. Navy term for a full, bright moon.."one which would help the older pilots of commander rank and weakening eyesight land safely on the deck at night."

compass rose — a circle graduated in degrees marked on the ground at an airport or printed on a **chart** and used as a reference to either true or magnetic direction.

compressibility — buildup of shock waves ahead of and throughout the airframe at very high speeds, producing **buffeting**. Meeting or approaching the sound barrier generated the theory that there was a maximum of resistance from air, before "breaking" (through) the sound barrier.

computational fluid dynamics — modern name for **wind tunnel** tests.

Conc, Conckers — refers to the Aerospatial Concorde SST and frequent fliers. "None of this deters loyalists, who call the plane the Conc and are known as Conckers." —Toronto *Globe & Mail*, March 13, 1996.

Condor Legion — German expeditionary air force fighting for General Francisco Franco in the Spanish Civil War.

coned — caught in a group of searchlights. WWII. —Doug Sample CD, Canadian Branch, Yorkshire Air Museum. The Canadian National Film Board film "The Valour and the Horror" describes the terrifying sequence: one searchlight would find a bomber, at which point many other lights would lock onto it, illuminating it in a bright cone; it would be a highly visible target. The pilot's only option was to dive to try to escape the cone.

conga line — a group of snowplows working in formation to clear a runway. — the movie *Airport*, 1970.

conk — the sudden stopping of an engine during flight due to a malfunction. From general slang, "conk out."

connect the dots — among "stewardesses," "flight plan," according to John Davis, *Buzzwords*.

CONs — contingent passengers, flight attendants' term for airline employees flying in otherwise empty seats. To be distinguished from **deadheading**, in which the employees have a reserved seat because they are working or commuting to or from work. Also **non-rev**.

conservatory — a power-operated, glass-enclosed machine-gun turret.

contact — term used by the mechanic as he was about to swing the prop to start an early aircraft engine, meaning "Is ignition on?" Pilot would answer "contact" to confirm. The term "no contact" was equally important to the prop-swinger if he first had to rotate the engine a few times to prime the fuel and loosen the oil. A miscommunication at this stage could lead to dire consequences for a prop-swinger unprepared for an engine to spring suddenly to life. See **prop, swinging the prop**.

contact VFR — flying in visual contact with the ground, being able to see it.

contour flying — flying close to the ground; **hedgehopping**. — Williams, *Casey Jones Cyclopedia of Aviation Terms*. Low flying. See **balls**.

contrail — condensation of "condensation trail," a visible line in the sky caused by hot engine gases forced into contact with cold air at altitude. Also **vapor trail**.

control authority — what the rudder and throttle, properly used and co-ordinated, are supposed to give the pilot, so that he may do his job right.

control column — See **joy stick, control yoke**.

control yoke — what newer aircraft have instead of a **joystick**. But the control yoke, big and mid-cockpit on the new Boeing 777, is a version of the old system. —*International Herald Tribune*, September 5, 1994. See also **pipe, yoke, stick, antlers**.

conventional aircraft — aircraft with a tail wheel. To be distinguished from aircraft with nose wheel/"tricycle" landing gear. See also **taildragger, tricycle**.

converter — any loud, inefficient airplane because "they convert fuel to noise." Specifically said of the Cessna T-37 jet trainer.

convertiplane — hybrid design incorporating rotary wings, fixed wings, tail-booms, and pusher engine, such as on the McDonnell V-1 V/STOL research aircraft. —*The Naval Air War in Viet Nam.*

cookie — in WWII, a 4000-pound or larger bomb. See also **blockbuster, tallboy.** —from Doug Sample CD, Yorkshire Air Museum, Canadian Branch.

cool — expression creeping into VHF vocabulary in place of **roger** or **wilco**. Heard mainly in Southern California.

cooler — guard room or detention room, in WWII, all branches of the military. It has entered general slang, for "jail." See also **glasshouse, digger.**

Co-op mix — jocular and cynical term among insiders at Short Brothers (aircraft, missile, armaments manufacturer, Belfast) for a particular blend of diesel fuel and fertilizer, which when it burns produces incredible power in a short time to move missiles fast.

copter — slang for helicopter, **chopper, egg beater, sling wing, whirlybird.**

copy — radio term used to indicate message received and understood.

coriolus — electrostatic discharge on aircraft windshield, wings, etc., popularly known as the spectacular "St. Elmo's fire."

corkscrew — 1. evasive action taken by a bomber to avoid fighter attack. WWII. —Doug Sample CD, Canadian Branch, Yorkshire Air Museum. 2. the manner in which an accident aircraft falls from the sky. "We saw the starboard wing separate and then part of the tail, and then the rest corkscrewed to earth." — quoted in *Flight International Magazine.*

corral — in naval aviation, the aircraft parking area on the carrier flight deck.

count the insulators — "to fly low." — Rose, *A Thesaurus of Slang.* See **Low flying, balls.**

crab — fly sideways, sort of, to skid on purpose. To slip or fly sideways to compensate for wind drift. **Sideslip.** See also **crosswind landing** and **ground loop.**

crabs — "Air Force (nickname used by other services; UK source)" — From John and Adele Algeo, "Among the New Words: Gulf War Glossary," *American Speech* (Winter 1991). The OED supplement has it as "midshipman."

cranked wing — see **bent wing**. A **gull** or **inverted gull wing**.

crash — **auger in,** creating a **smoking hole,** etc. One of the earliest famous ones was at Bournemouth, when the Hon. Charles Rolls, with his engineer Henry Royce, creator of the Rolls-Royce auto and engine company, and a very early flying enthusiast, after becoming the first British flyer to cross the Channel by air and the first of any nationality to fly both ways, died by....augering in.

crashbait — among "stewardesses," "frequent traveler," according to John Davis, *Buzzwords*.

crash bay — shed where emergency vehicles were parked, WWII military aviation. —Doug Sample CD, Canadian Branch, Yorkshire Air Museum.

crash patch — among "stewardesses," slang term for the runway, according to John Davis, *Buzzwords*.

CRASS — said to be descriptive of Gulf War (mostly aviation) language: Cryptic References, Acronyms, Slang and Shorthand. — From John and Adele Algeo, "Among the New Words: Gulf War Glossary," *American Speech* (Winter 1991).

crate — affectionate, older slang for "airplane." According to Dugal in *Readings on Aviation* (1929) "A crate is an airplane. This is a favorite slang expression among soldiers and commercial flyers." See **aircraft, kite, bird**.

creep-back — the tendency of bombers in a stream, WWII, to drop early, thus impact points move steadily "backwards" from the intended impact zone.

crewed-up — during WWII bomber operations in England, to have completed the process of forming up, voluntarily, the bomber crew from the assigned available men. Complete crew of the **Halibag**, for example, included pilot, navigator, bomb aimer, wireless operator, and air gunners. If a bomber did not successfully "crew up" on the voluntary basis, command would assign men to fly together. —from Doug Sample CD, Yorkshire Air Museum, Canadian Branch.

critical wing — term used to describe a thin, narrow wing, designed to offer the minimum amount of surface necessary to provide lift. Efficient but unstable, used in modern high-speed aircraft. See **coffin corner.**

crosscheck — a verbal reminder from the cockpit to the cabin crew on some airliners to verify whether the ejection slides at emergency exits have been "armed," i.e. turned

on so that they are ready for explosive emergency deployment.

cross country — flying to at least three places twenty or more miles apart.

crossed controls — said of the skidding technique using opposite rudder and aileron. How to **crab**.

crosswind landing — to give the aircraft more lift, most landings and takeoffs are into the wind. If the wind is cross-wise to the runway, the pilot must **crab**, and if the craft is a **taildragger** and the rear wheel happens to be turned sideways upon landing, the plane might do a **ground loop**.

crow hop — practice procedure in which a student lifts the airplane off the ground for a second or two, then lets it bounce back down, during the early stages of training.

cryoplane — aircraft burning hydrogen (cryogenic fuel). —*ILS News* (Berlin Air Show), June 16 1992.

Cuban eight — an aerobatic manoeuver in which an airplane completes three quarters of a loop, rolls over and completes the loop in the opposite direction, resulting in a vertical figure eight. See **eight**.

cube — the available space in an aircraft.

cubed-out — an aircraft in which all the available space has been filled.

cul-de-sac — taxiway with an obstruction or barrier at one end.

cumulogranite — "an aviator's term for foggy mountain meteorology, particularly coastal. It's desirable to avoid such hard-centered cloud when flying through **clag**." —Parkin, *WetCoast Words*. See also **stuffed cloud**.

Cunni-Lingus — impolite nickname for Aer Lingus, the Irish airline. Also said of an imaginary merging of Continental Airlines and Aer Lingus.

Curly —common "stewardess" name for the navigator, according to John Davis, *Buzzwords*. See also **Larry, Moe**.

cut-and-try — engineer's term for trial-and-error.

Dachshund bellies — 1000-litre long-range fuel tanks slung beneath Messerschmitt Bf-110 fighters by Luftwaffe in WWII. "These clumsy tanks were notoriously unreliable and this one did not fall away before a bullet touched off the vapour. His Bf-110 exploded like a bomb." —Deighton, *Fighter*.

daisy chain — in naval aviation this refers to a group of aircraft descending in single file from altitude to make an instrument approach or a carrier-controlled approach.

daisy cutter — a three-point landing.

daisy cutter bomb — a 10,000 lb. bomb used in Viet Nam to create instant helicopter landing zones by flattening all vegetation and most structures within an area several hundred yards in diameter. — Mersky and Polmar, *The Naval Air War in Viet Nam.*

DALPO — death case passenger, in ticket agent parlance.

DALPOSE — "Do All Possible to Assist." —flight attendant and ticket agent code, from John Cavill. **DALPO**, however, means "death case."

Dam Busters, the — nickname of 617 Squadron, RAF Bomber Command, WWII, famed for undertaking highly specialized operations against difficult targets. At the squadron's inaugural dinner, the printer erred and labelled the menu "The Damn Busters." See also **bouncing bomb, skip bomb.**

dash modes, dash speed — "There was a Grand Canyon-size gulf between designing an airplane like the F-104 that could kick in its afterburners on takeoffs and in dash modes lasting a minute or two, and designing an airplane whose 'normal' cruising speed was nearly twice as fast as the fastest fighter's dash speed." — Rich, *Skunk Works.* See **burst speed.**

deadbeat — RAF WWII term for non-flying personnel. Also known as **penguin, ground pounder, ground hog, kiwi.**

deadheading — see **CONs.** Paul Dickson defines it thus: "for a crew member to fly as a passenger, either to return home or to catch a flight to which they are assigned. One is only considered to be deadheading when in uniform, so going on vacation in civvies does not count." —*Slang!* Also **non-rev.**

dead reckoning — "ded. reckoning," "deduced reckoning," the navigational scheme, by which originally ships, and later aircraft, would attempt to locate themselves by reasoning from a known position and factoring in speed, direction, time elapsed, etc. to estimate current position. In zero visibility, when the electronic aids are down, "you can't get out of that jam without it." "You better reckon right or I reckon you're dead!"

dead stick landing — From *What's the Gen,* by HW (1942 pamphlet of slang and cartoons from the RAF): "dead stick: a propeller stopped in the air." Power off, landing on control surfaces, like a glider. Not what you'd choose to do, but useful when making a clandestine descent into an unlighted or disused field. You chop the engine on final approach and "dead stick" it silently in. More frequently, the glide in after engine failure. Still used on non-prop, jet, aircraft for a "no power approach."

The term was subsequently adopted by American teenage drivers in the form "**dead-sticking** it in" — "when you had borrowed your Dad's car, with the solemn promise to have it home by midnight, say, and you were very late, *and* you could see by the lights that he had probably fallen asleep, so there was an outside chance that you could shave a bit off the penalty by claiming you hadn't been all *that* late, you would cut off the engine half a block before reaching the driveway and try to coast it in. You'd say, 'I had no choice but to dead-stick it in!'" — Poteet and Poteet, *Car & Motorcycle Slang*. See also **Gimli glider.** Another famous dead-stick landing was made on the mud banks of the Thames River in London by a Bristol Britannia.

Also, a **joy-stick landing; one burning, one turning.**

death card — jocular name for the "safety procedures" card inserted in the pocket on the back of the seat in front of you on an airliner, as applied by the personnel who service the cabin.

death ray — term for the use of a lock-on to target with missile-guidance radar, during the Gulf War. —U. S. News & World Report, *Triumph without Victory*, 1992.

debark — to descend from an airplane. Borrowed from boat talk. See **embark.**

debriefing — when the Intelligence Officer (IO) questions crew to ferret out **ops** information/details. See also **briefing, making their numbers.**

decelerons — combination ailerons/airbrakes to assist in rapid deceleration, such as found on the Northrop F-89 Scorpion.

decision height — the published minimum altitude (varies with different airports) at which a pilot must decide whether to land or to execute a missed approach.

deck dub — in naval aviation this refers to an aircraft ready for flight but unable to be launched.

deep black — "Among the services, the Navy was the most active in running 'deep black' programs, ...but 'Your stealth fighter is the first black program the Air Force has ever run. Security is paramount.'" — Ben Rich, *Skunk Works*.

deflected slipstream — principle which involves the use of large diameter propellers which bathe virtually the entire wing in their slipstream. Large trailing-edge flaps deflect the slipstream downwards, producing lift. One example is the Breguet 941.

de-icing boots — undulating rubber surfaces on leading edges designed to crack and shed ice forming during flight.

Delta — Don't Even Leave The Airport, an irreverent acronym for Delta Air Lines. See also **Destroy Every Living Tree Around**. 2. In naval aviation this refers to a specified area for carrier aircraft waiting to land. Also called the **dog circle**.

delta fins — a pair of stabilizing fins at the rear of the fuselage with a downward slant and below the horizontal stabilizer, built into the design of modern Learjets and the Piaggio Avanti, to help in stalls. —Bill Stratton.

Delta Force strike, Delta Force-style strike — a covert **surgical strike**. Analogous to **Alpha strike** which refers to massiveness and coordination among several forces; here the emphasis is on covertness. "The squadron received top secret orders directly from Caspar Weinberger, Reagan's defense secretary, to be prepared for a Delta Force-style nighttime strike against Libya's Muammar Kaddafi's headquarters in Tripoli." —Ben Rich, *Skunk Works*. See also **silver bullet**.

delta wing — a triangular-shaped wing.

demolition derby — Joe Gadget, pseudonymous Canadair employee, reports that this term is used to describe stress testing an airplane to find the life expectancy of each of its components.

dense at the fence — rhyming slang, among "stewardesses," for a situation where many planes are being held at the gate, according to John Davis, *Buzzwords*.

Derry turn — RAF manoeuvre in which one rolls below enemy and reverses turn to come up under his belly. —Capt. J. A. Kent, *One of the Few*.

Desert Cat — "Member of the Canadian Air Force in the Gulf region" — From John and Adele Algeo, "Among the New Words: Gulf War Glossary," *American Speech* (Winter 1991).

designed by morons for midgets — said of a too-small cockpit. —McVicar, *Grass Runway*.

Ideal aircraft is **designed by pessimists, built by perfectionists, and flown by cowards.** — engineer's and old flying enthusiast's saying.

Destroy Every Living Tree Around — according to an employee of Delta Airlines, at one point, after a couple of crashes, the Federal Aviation Administration decided that some maintenance records were missing and ordered that such records be kept in multiple copies; the airline itself also followed with an order for increased multiple retention of maintenance records. The result was such a heavy paperwork burden that the employees themselves supplied the explanation for the "acronym" D E L T A.

de Wilde — a type of machine gun ammunition used by the RAF during WWII which combined the qualities of tracer and explosive with a solid armour-piercing core. "...bullets made small dancing yellow flames as they exploded against the enemy fighter. This made it easy to check one's aim." —Deighton, *Fighter*. See also **Buckingham**.

DF — direction finder. Usually a radio beam. The instrument in the cockpit is tuned to the beam and a needle indicator points to the direction from which the signal is being transmitted, allowing flight "to" or "from." Later known as Automatic Direction Finder (ADF).

diamond landing gear — configuration with tandem mainwheels on the centerline and outriggers on the wings.

dicing missions — "low altitude tac recce" missions. —Bob Stevens, *"There I was."*

digger — WWII RAF slang for the guardhouse, place to serve a minor punishment. See also **glasshouse**. —Doug Sample CD, Canadian Branch, Yorkshire Air Museum.

dihedral — the upward inclination of the wings. See **anhedral, gull wing, inverted gull wing**.

dipsydoodle — dive and pull manoever used by Lockheed SR-71 Blackbird pilots to help the airplane through Mach 1.

dirty — an airplane flying with gear, flaps, spoilers, or everything down, extended. See also **clean**. May also be said of weather with poor visibility, low ceiling.

dirty bird — In the development of the U-2 spy plane at Lockheed, "[radar-absorbing] paint lowered the radar cross section by one order of magnitude, so we decided to give it a try. We called these specially painted airplanes 'dirty birds.'" —Ben Rich, *Skunk Works*. "In July 1957 a cargo plane brought the first so-called dirty bird to our base in Turkey. It was covered with a plastic material and had two sets of piano wire strung from either side of its nose to a set of poles sticking out of the wings. The wires were to scatter radar beams while the paint was to absorb other frequencies....I had never before risked flying an airplane wired like a guitar." —James Cherbonneaux, quoted in Ben Rich, *Skunk Works*.

dirty dive — a dive using air brakes, spoilers, flaps, undercarriage, etc., to help keep it slow. "I was set for my dirty dive." —Gord Squires, "Joy Riding in an F-86", *Aircraft* (April-May 1990).

dispersal — area where aircraft were parked (WWII). See also **pan**.

ditch — crash-land in water. "Ditched in the drink." —Bruce Callander, "Jargon of the Skies," *Air Force Magazine* (October, 1992).

dive — Jocularly, the pilots' lounge. Known at certain air bases and airports as the "Auger Inn."

dive bombing — target approached in steep dive, bomb dives to target as aircraft pulls up.

dive brakes — flap or spoiler-like devices to help slow or stabilize the aircraft during a dive. Also **airbrakes, speedbrakes**.

diving for the deck — "Jake pulled off a handful of power, moved the stick forward a smidgen, then pulled it aft as he shoved the power back on. This manoeuvre violated every rule in the book — it was called "diving for the deck," but it was a sure way to get aboard when you had to." —Coonts, *Flight of the Intruder*.

DOA — dead on airplane, among "stewardesses," according to John Davis, *Buzzwords*.

Dobbas — Luftwaffe WWII term for "collapsible freight and equipment container. This aerofoil-shaped container was intended to carry spares when units were moving from one operational field to another, or urgently needed freight had to be moved fast, and was slung beneath the Ju-87 main undercarriage members, being attached to the central bomb rack." —William Green, *War Planes of the Second World War*.

doctor in a Bonanza — usually a derogatory term describing a non-professional flyer, but to remote communities visited by **circuit riders**, often a compliment. See **fork-tailed doctor killer**.

dodo — "a pilot trainee who has not yet flown solo." —Woodford Agee Heflin, ed. *The United States Air Force Dictionary*. See **red streamer**. 2. According to Assen

Jordanoff, *Jordanoff's Illustrated Aviation Dictionary*, a member of the air force who does not use his wings because of a desk job or by personal preference.

dog circle — in naval aviation this refers to a specified area for carrier aircraft waiting to land. Also called **Delta**.

dogfighting — see **ACM**.

dog house[s] — "small storage compartments throughout an aircraft used primarily for safety equipment." —American Airlines training materials.

dog-leg — 1. a course flown at right angle to the intended course so as to use up time when ahead of schedule. It was important for the **stream** of bombers in WWII to arrive at the aiming point for a target in coordination. —Douglas Sample CD, Canadian Branch, Yorkshire Air Museum. 2. Term applied to the German radio guidance system that WWII bombers followed to targets in England. Translated from the German name "knickebein," or "dog-leg." British code name was **headache** and the jamming device eventually used to counter it was code-named **Aspirin**. —Deighton, *Fighter*.

Dog Mushers Society — "Qualifications for entrance require that a pilot shall have mushed dogs at least one hundred miles. Membership in the society entitles the pilot to let his whiskers grow." Formed by pilots flying for Western Canada Airways on Northern operations, 1927. The usefulness of this skill to an early flyer in the remote North is obvious if you think about what you do after you survive the crash in an isolated snow-covered area.

dogs — passengers. — Paul Dickson, *Slang!* See also **self-loading cargo, talking ballast, SOBs** and **PAX**.

dog-tooth — (Brit.)step in a wing leading edge where chord (breadth of wing from front to rear) increases. Also **saw-tooth**, useful in maintaining stability at various speeds.

domestic — refers to flights within the country, as opposed to **transat** (transatlantic) or other international flights.

doodlebug — 1) type of bomb used by RAF in WWII. 2) British nickname for German V-1 rocket, also known as **buzz bomb**.

doolie — U.S. Air Force Academy cadet. — Paul Dickson, *Slang!*

Doolittle raid — "an air raid (flown) by North American B-25s (Mitchells) against Tokyo on 18 April 1942, led by Lt. Col. James H. Doolittle." —Woodford Agee

Heflin, ed. *The United States Air Force Dictionary.*

dope — "liquid which shrinks and tightens the fabric" (canvas, in early aircraft). —Howard N. Rose, *A Thesaurus of Slang.* 2. information, poop, **gen**.

Doppler — the effect produced by sound from an approaching or departing aircraft, used to find relative position through Doppler instrumentation. Also used in radar parlance.

"Originally associated with sound, it is typified by the phenomenon of the approaching, passing, receding train with a whistle blowing. The steady pitch of the approaching whistle suddenly drops as the train passes, to a new, lower note, which is maintained. Stationary, the whistle emits sound waves per second (frequency) appropriate to the pitch, that come at you at the speed of sound. Approaching, the sound comes at you at the speed of sound plus the speed of the train, so more "waves" (higher frequency) are coming at you in the same time; and you hear this as a higher pitch. For a split second, as it passes, you hear the true pitch. As it recedes, the sound of the whistle comes back at you at the speed of sound less the speed of the train, hence, fewer waves..., lower frequency...., lower pitch. This same Doppler effect works with radar waves to assess aircraft speed relative to the ground to establish position, relative speeds - of other aircraft, or relative speeds in general; like yours versus the cop with the radar gun!" —anon. British Canadair technical writer informant.

double bubble — name given a type of fuselage shaped like a figure 8, usually containing two decks, as in the Boeing 377 Stratocruiser. Also called a **double decker.**

double decker — usually a biplane, but more recently, a large double-decked transport, such as the 747.

double nut — name given by USN Viet Nam pilots to the "00" numerical markings on fighters flown by the air wing commander, always called CAG for "commander air group" despite redesignation in 1962. The term suggests that the commander has **balls.**

down and dirty — gear and flaps down. See **dirty**. An adaptation of slang from poker.

down and locked — refers to landing gear in the landing position. Often asked by the air traffic controller of the pilot when issuing landing clearance—whether the landing gear is ...... See **three greens, up and locked.**

down and welded — fixed landing gear. —from Paul Turk.

down gripes — serious aircraft deficiencies that need immediate attention, often grounding the aircraft until repairs are made. —Coonts, *Flight of the Intruder*. See also **up gripes**.

downwind landing — See **Chinese landing**.

downwind leg — in manoeuvering for the **final approach** to the runway and landing, it is sometimes necessary to fly downwind to a certain point, then make a **base leg** flight at right angles, finally turning directly into the wind for the final approach and landing. See **circuit**. A jocular definition goes, when a girl is standing sideways to the wind, skirt will be lower on this leg.

doying — said of "a new pilot coming in to land for the first time." —W. H. *What's the Gen?* See **red streamer**.

drag chute — a parachute deployed from the rear of an aircraft to help slow it down on landing. See **drogue**.

dragging — flying low over a landing ground not regularly used, to determine the condition of the surface and the best approach for landing.

the right **drill** — the correct method. WWII.

drip torch — "In 1973, John Muraro of the Canadian Forestry Service at Prince George BC invented a helicopter-suspended fuel barrel to ignite logging slash. A 60/40 mixture of diesel oil and gasoline is ignited electrically, then dribbled by the pilot while in flight over clearcuts. Such site preparation increases the survival and growth of tree seedlings. Called a helitorch in other provinces." —Tom Parkin, *WetCoast Words*.

driver — though most widely known as the **pilot,** the person in charge of operating an aircraft is also known, especially to other pilots and when referring to him or herself, as a **driver**. Specialized uses include "transport driver," "bomber driver," "jet driver," "fighter driver," etc.

driving the train — leading two or more squadrons into battle.

drogue — a word from boat talk, where it means "sea anchor," used in plane talk to describe the connecting equipment coming from a tanker for inflight refueling ("put the probe in the drogue") and also for a parachute used to slow a landing aircraft after **touch down** on a short runway: **drag chute**.

2. In HW's *What's the Gen?*, a 1942 pamphlet of RAF cartoons and slang, "droque" is glossed as "a target carried by an aircraft, on the

end of a long cable, for other aircraft to shoot at."

drop out — "eject," **punch out.** —Murray, "Language of Naval Fighter Pilots."

drop tanks — jettisonable underwing fuel tanks. (Empty them, then throw them away).

dry sump — racing modification. Oil is pumped out of the oil pan faster than the lubrication system pumps it in, via the crank and cam and valve gear, thus allowing crankshaft to achieve higher rpm, more power, because the crankshaft is not "dragging" through the oil. Originally developed for airplanes, because with the engine sometimes flying upside down, etc., you couldn't rely on the reservoir of oil to flow out by gravity. For other automobile terms and bits of technology derived from the airplane, see also **Dzus fasteners, supercharging,** and think "fly low."

DSL — "Dangerous Sperm Load, a condition resulting from being in the field and away from your sweetie for more than twenty-four hours." Hans Halberstadt, *Airborne: Assault from the Sky.*

duck and cat navigation in fog — when ILS is out, find a flying duck and follow it down; keep a cat standing up straight in the cockpit, to show true direction upwards and to overcome disorienting effects of vertigo.

duct rumble — In U-2 spy plane testing, Lockheed "test pilots began reporting 'duct rumble' at fifty thousand feet, describing the sensation as driving down a deeply rutted road on four uneven tires. In an airplane as fragile as the U-2, such severe shaking was a serious problem. The cause was flying at a slant so that more air was entering one of the twin air-intake ducts than the other....Dave helped me design a splitter to enhance more even airflow....Pratt & Whitney finally solved the problem completely a year or so later by revising the fuel control for a better match of air and fuel into the engine." — Ben Rich, *Skunk Works.*

duff — British word for "unflyable weather." —McVicar. Also in W. H. *What's the Gen?* See **clag.** "Duff days" were days when the weather was "unsuitable for flying," according to Monty Berger, *Invasions without Tears: The Story of Canada's Top-scoring Spitfire Wing in Europe during the Second World War.* Also used for wrong or incorrect data or information.

duff bird — an aircraft which is **unserviceable**, or barely flyable.

duffle bag — a radar-reflective ground sensor detectable by airborne radar as a way of marking ground locations. Viet Nam era. —Ray Seale.

Dumbo — U.S. Navy and Marine term for rescue missions in the Pacific Theatre, World War II. "The dumbo rescue missions were universally boring because we had to fly our fighters dead slow to remain on station with twin-engine PBY Flying Boats or single-engine Grumman Ducks, neither of which could fly straight and level at over 125 mph. On the other hand, the payoff for these missions was usually the rescue of a downed aviator." —Col. R. Bruce Porter, *Ace! A Marine Night-Fighter Pilot in World War II.*

dummy run — bomb aimer having missed his chance to release bombs on target, aircraft has to go around again and try another run-up to the target. A source of dis-ease among crew, as it exposes them to more **flak** and increases likelihood of being hit. WWII. —Doug Sample CD, Canadian Branch, Yorkshire Air Museum. The term "go around" usually refers to a landing situation, but to bomber crews often meant a second approach to the target.

dungeon — common "stewardess" slang for the cockpit, according to John Davis, *Buzzwords.*

dupeoffed — "deployed off, 'bumped,' because of oversale of seats." —from John Cavill. Means you're looking for **FIRAV** space.

dustbin — (Brit.) gun station projecting below fuselage of aircraft. See **ashcan**.

Dutch roll — the oscillating motion of an aircraft involving rolling, yawing and sideslipping. The term is used because of the likeness to the characteristic rhythm of an ice skater. —used by David Hughes, Pilot Report in *Aviation Week and Space Technology* (Farnborough issue, September 7, 1992). For a somewhat different definition see **roll**.

duty runway — the runway in use at the time. See **active**.

Dzus fasteners — a sort of quick-release fastener or clip, using a flush-fitted key- or coin-turnable latch, for, e.g. hood and fender skirts. Designed for minimum wind resistance, and first developed for airplanes.

Eagle of Africa, the — nickname given German fighter pilot Hauptmann Hans-Joachim Marseille, who claimed 158 victories flying Messerschmitt Bf-109s, WWII.

earache — among "stewardesses," "a talkative passenger," according to John Davis, *Buzzwords*. For other derogatory terms for difficult clients, see **claw, Etna, poker, ten-pins, faucet, sweater, jump-seat sniffer.** Flight attendants also have slang terms for passengers based on national origin: "Cannoli" for an Italian, "kamikaze" for Japanese, "Stuka" for German, "bullet hole" for Indian, and "Goth" and "car bomb" for Irish.

earthquake bomb — in WWII military aviation, a 22,000 lb. bomb, also known as the **Grand Slam.** See also **blockbuster, cookie, doodlebug, tallboy.** —Doug Sample CD, Canadian Branch, Yorkshire Air Museum.

eat and run — term used in Viet Nam for "rearm and refuel." See **turn 'em and burn 'em.**

EATCHIP — European **ATC** Harmonisation and Integration Programme. Project aimed at creating a seamless Europe-wide Automatic Traffic Control system to replace the present motley collection of individual national systems, integrating over 50 control centres, using more than 30 different, sometimes incompatible, ATC systems, involving many different computer operating systems and "languages." See **Single Sky for Europe, open skies.**

economy class syndrome — "Frequent fliers, take note: a little fidgeting can ward off blood clots like the type that lodged in Dan Quayle's lungs after he rode on one too many airplanes....A few people get clots from simply sitting too still—lying in bed after surgery or sitting in a cramped airplane seat for hours, a phenomenon sometimes called 'economy-class syndrome.'" —Chicago *Tribune.* "Among the New Words, *American Speech* (Summer, 1995). See also **coach-seat clot**

E & E net — "Escape and Evasion route, a covertly established route network for rescuing fliers downed behind enemy lines." —From John and Adele Algeo, "Among the New Words: Gulf War Glossary," *American Speech* (Winter 1991).

egg — affectionate name for "bomb." See also **vertical egg,** a name for a type of manoeuvre.

egg beater — slang for helicopter, **chopper, copter, sling wing, whirlybird.**

ego containers — business jets which are converted larger aircraft, such as Donald Trump's Boeing 727 or international arms dealer Adnan Khashoggi's leased 707, according to Maureen Orth, in "Jet Compulsion," *Vanity Fair* (December 1995).

eight — any of a number of flight manoeuvres in which an airplane follows a horizontal flight path in the form of a figure eight, the only exception being the lazy eight in which the nose of the plane describes a vertical eight: Eight across a road - an elementary flight manoeuvre describing two loops of equal diameter, one on each side of a road in both medium and steep turns; Four bank eight - similar to eight across a road but with each loop formed by two 45-degree turns connected by a straight flight portion; Eight on pylons - an advanced flight manoeuvre forming the figure eight around two pylons with a plane's lateral axis theoretically pivoting on each pylon in turn; Eight around pylons - an intermediate flight manoeuvre forming the figure eight around two pylons at right angles to the wind direction, 270 degrees of each loop being a true circle; Cuban eight - a flight manoeuvre comprising three quarters of a normal loop, a half roll, three quarters of a normal loop, and another half roll. —Henry L. Williams, *Casey Jones Cyclopedia of Aviation Terms*.

eight hours, bottle to throttle — Federal Aviation Administration (U. S.) rule for pilots; airlines have a 12-hour rule. According to Ron Cosper, St. Mary's University (Halifax), naval pilots have a separate one: "no alcohol within 25 feet of the aircraft."

eject — to use the mechanism that propels the pilot, with parachute and often seat, out of the cockpit in emergencies. See **bang out**.

ejector seat — aircrew member's seat mounted on an explosive charge that propels the crew member out of the aircraft in an emergency. If this is done at low altitude while flying upside down, the seat and crew member end up 25 feet down in the mud. Also called "ejection seat."

electronic countermeasure aircraft — "planes equipped with ECM (electronic countermeasures) to counteract radar and other electronic devices." —From John and Adele Algeo, "Among the New Words: Gulf War Glossary," *American Speech* (Winter 1991).

element — two or three aircraft. In increasing order of hierarchy and numbers, it's **element, flight, squadron, wing.**

element leader — "Pilot of a plane that another pilot, the **wingman**, follows." — From John and Adele Algeo, "Among the New Words: Gulf War Glossary," *American Speech* (Winter 1991).

elephant trunk — WWII British expression for the flexible tube from the oxygen tank to the face mask.

elevator — jocularly, device that raises the runway thus preventing pilots from "dropping it in."

embark — to start on a journey, to board an airplane. As with **debark**, borrowed from boat talk.

empennage — tail assembly, rudder and horizontal stabilizers. "How far can wing and fuselage be stretched? These are empennage considerations." In this case the empennage can actually be smaller, more "close-coupled," because the extension gives the tail surfaces more purchase on the air.

The word "empennage", along with a number of other key terms in aircraft terminology (**aileron, canard, barbette, nacelle, pitot tube, chandelle, fuselage**), comes from French, pointing to the key early role of France in developing aviation. Some say that they do have in fact a solid claim to have been the first, ahead of Otto Lilienthal, a German, and the American Wright brothers. There's even a Russian claim, someone's having towed a glider behind a coach! To some degree, of course, the claim depends on definitions of "manned flight," which need to be precise. See also **Father of Aviation, Icarus**.

empty kitchen — "Name given to female pilots by males—presumably from the fact that the woman is in the cockpit of a plane rather than in her house." — Paul Dickson, *Slang!* See also **box office**. An unprogressive expression.

engine — in plane talk it's always "the engine" or the **powerplant**, never "the motor," though before WWII "motor" was frequently heard along with "engine."

Take the cylinder out of my kidneys,

The connecting rod out of my brain, my brain,

From the small of my back take the crankshaft

And assemble the engine again.

— RFC Mess song - WWI

engine surge — phenomenon common to the Aerospaciale Concorde supersonic jetliner; throttling back from Mach 2 causes the engines to produce a 'forward-firing backfire' due to the sudden breakdown of airflow through the engines.

engine wash — an alternative term for **wake turbulence**. But see also **prop wash**.

envelopes — the outer limits (boundary lines) of an aircraft's capabilities, such as speed, stall speed, range, ceiling, etc.

Environmental Air Force — Headquartered in Philadelphia, PA with offices in Alameda, CA and Brunswick, ME, this is a group of volunteer pilots who fly over areas of environmental concern to photograph, map, survey and study, making their information available to the proper authorities for corrective or preventive action.

equipment — "the plane. If you are told that there has been a slight delay in the arrival of the equipment, it means your plane is late." — Paul Dickson, *Slang!*

erks — aircraftsmen (Brit.) Comes from the Liverpool accent pronouncing the word, in which "airc" is pronounced to rhyme with "perk," as in "Aircraftsman 2 - second class," ["erkraaftsmun"], LAC ("leading aircraftsman"), or SAC ("Senior aircraftsman"), "er-craftsman," shortened to "erk."

EROPS — Extended Range Operations - applied to commercial twin-engine overseas flights, jocularly called Engines Running or Passengers Swimming. Earlier known as **ETOPS**.

ErrorMexico — slang term for AeroMexico, during "late 1988, when the airline was government-owned, bankrupt....." Sam Dillon, "Flight into Nowhere," New York *Times* (August 10, 1995).

escape photo — series of photos aircrew carried (WWII) on **ops** to be used for false ID photos by underground in the event one was shot down. —Doug Sample CD, Canadian Branch, Yorkshire Air Museum.

ETA — estimated time of arrival.

Etna — among "stewardesses," slang term for a smoker, according to John Davis, *Buzzwords*. See also **poker, faucet, sweater, claw, tenpins.**

ETOPS — Extended Range Twin-Engine Overwater Operations (first commercial flight El Al Boeing 767 Montréal-Tel Aviv, March 26, 1984). Engines Turning or Passengers Swimming! "For those who can swim, welcome to Bermuda. For those who can't, thank you for flying Air Canada." Also known as **EROPS** - Extended Range Operations: Engines Running or Passengers Swimming.

evader — airman who has been shot down, parachuted out or otherwise survived, in enemy territory, and who hides out, evading capture which would make him a POW. See also **Walker's Club, Caterpillar Club, Flying Boot Club.**

execjet — see **bizjet.**

eyebrow window — small window above the windshield for upward/forward visibility. See **lookdown window.**

eye in the sky — police or traffic helicopter, also refers to unmanned airborne surveillance systems (carried on Unmanned Air Vehicles). See **telecopter**.

FAA — Federal Aviation Agency (U. S.) Government regulatory agency in charge of virtually all aspects of flight.

faceting — "creating a three-dimensional airplane design out of a collection of flat sheets or panels, similar to cutting a diamond into sharp-edged slices." —Ben Rich, *Skunk Works*. See **Hopeless Diamond** in "Plane Names."

fail-safe — back-up systems that kick in should the primary system fail. Alternate paths for loads. Multiple redundancy, a basic principle of aircraft design and outfitting aims to supply, as far as possible, *bypass systems*, "alternate means of controlling an aircraft under certain emergency conditions in flight." — Beaudoin, *Walking on Air*.

fan — the propeller (WWII). —Doug Sample CD, Canadian Branch, Yorkshire Air Museum. Modern usage includes "fan jets," which are high-bypass-ratio jet engines.

fan marker — an electronically defined point in an instrument approach. See also **outer marker, middle marker**.... —from Paul Turk.

FANS — Future Air Navigation System, a planned worldwide scheme for satellite control of air traffic. Also known as **GPS**, Global Positioning System.

farm — the area beyond the runway. A place where it is possible to **buy the farm**.

Farman — Henri and Maurice, early French/English flyers, associated with the exploration of Africa and the creation of Air Inter.

Farnborough — major international air show, every two years (1992, 1994 etc.) held at Farnborough, England, southeast of London. Originally the site of the Royal Aircraft Factory, which had overall responsibility for the design and construction of aircraft during WWI.

fart — among "stewardesses," a term for wind shear, according to John Davis, *Buzzwords*.

fast FAC; fast forward air controller — "F-16 stripped of most armament, used to scout an area for targets and mark them with phosphorous rockets, followed by heavily armed F-16s to destroy them." — From John and Adele Algeo, "Among the New Words: Gulf War

Glossary," *American Speech* (Winter 1991).

fast file — a system whereby a pilot files a flight plan by telephone or radio. Also see **air file**.

fast movers — jet fighter planes. (Viet Nam). Used with this sense in the movie "Forrest Gump" (1994).

fat air — among hang glider flyers, a description of preferable flying conditions. See **trash air**.

Father and Son — from the German *Vater und Sohn*, German **mistel** (**Mistletoe**) composite transport/air launch arrangement, a Messerschmitt Bf-109F over a Junkers Ju-88 A-4. See also **Beethoven**.

Father of Aviation — according to some British sources, not the Wright brothers or their father, but a British person, Sir George Cayley, who built a model glider in 1804, a glider big enough to carry a small boy in 1849, and a larger one, with his coachman as pilot, in 1852. See also commentary under **empennage**.

Father, Son, Holy Ghost — U.S. Navy mnemonic used during testing of the control stick or column as part of the pre-flight check, referring to left, right, forward, backward. Often the rudder pedals would then be checked, left, right to the sound of "a-men".

faucet — among "stewardesses," a nervous flyer, also known as a **sweater**, according to John Davis, *Buzzwords*.

FBO — Fixed Base Operator or Operation, a concession at an airport, with parking, fuel and servicing, coffee, etc., for private aircraft.

fear of flying — actually, this is fear of crashing. — Richard Gordon, former public relations executive, Shorts, Belfast.

feather — to rotate a variable pitch propeller on a dead engine so it offers less surface to the wind, preventing **windmill effect** or **windmilling**.

feet dry, feet wet — Carrier-based pilot's code for "I am now over land" and "I am now over water" in reporting progress toward destination. Also reported in use by USAF. See **coast-in point, coast-out point**.

fence — 1. at or near the end of the runway, as in, "try to get it in over the fence." 2. a projection along a wing's leading edge to prevent spanwise airflow.

fenestron — form of helicopter tail rotor consisting of many narrow blades in a short duct.

fercht oder getrocknet — from German, an expression used by Oswald Boelcke, German ace, during WWI, stating what he expected to be the alternatives for his own manner of death..."literally 'wet or dry' i.e. burned or mutilated to death. It was *fercht* following a collision in combat on 28 October 1916." — Alan Clark, *Aces High*.

Fidel — among "stewardesses," a slang term for a hijacker, according to John Davis, *Buzzwords*. The term derives from the large number of flights that were rerouted from North America to Cuba, during the days in the 1970s when hijackings were more numerous. See **skyjack**.

FIDO — 1. In WWII military aviation, in England, Fog Investigation and Dispersal Operation, involving gas flares along runway. —Doug Sample CD, Canadian Branch, Yorkshire Air Museum. Another definition for the anagram is Fog Instant Dispersal Operation.

2. "Fuck It, Drive On!" — "An abbreviated suggestion to overcome an obstacle to progress and get on with the program." Hans Halberstadt, *Airborne: Assault from the Sky*.

fifinella — a female **gremlin**, said to be more deadly than a male.

fifth freedom — under international treaties governing commercial aviation, the "right to take on, in the territory of another State, traffic destined for the territory of a third State, and to put down, in that territory, traffic coming from another State." —from Swissair's "ABCs of the Airline Industry." The others are, first, "right to fly across the territory of another State without landing"; second, "the right to make stops in the territory of another State for non-traffic purposes"; third, "the right to put down, in the territory of another State, traffic (passengers, cargo, and mail) taken on in the territory of the State whose nationality the aircraft possesses"; and fourth, "the right to take on, in the territory of another State, traffic destined for the territory of the State whose nationality the aircraft possesses." See also **cabotage**.

fighter affiliation — exercise used to sharpen air gunners' skills with practice attacks. —Doug Sample CD, Canadian Branch, Yorkshire Air Museum.

fighter pilot's oral rules — "Never throttle back in combat." "Beware the Hun in the sun." —Deighton, *Fighter*. "When in doubt, pull out." "Don't puke in your oxygen mask." —from the movie *Into the Sun*.

final or **final approach** — "lined up on runway heading and descending for a touchdown." "Final approach to the *"terminal"*?! To the wide-open ear, this bit of flight-attendant and air traffic controller parlance sounds as if it violates the rule against joking about disasters on airplanes, imposed relentlessly on almost all passengers. In fact, the term designates the last five to ten miles of the approach to the runway. The "final approach" comes after, sometimes, a **downwind leg** and a **base leg,** or it may be **straight in.** See **circuit.**

2. Jocularly, the last pass a pilot makes at a girl before giving up.

finger four — Allied term for the German "schwarm" formation, WWII. —Deighton, *Fighter.* See **v, vee.**

FIRAV — "first available," e.g. if you've been **dupeoffed,** you'll be looking for it. Ticket agent and flight attendant slang. —from John Cavill, former public relations for Air Canada eastern Canada region.

fire handle — handle in cockpit that cuts off fuel, hydraulics, and electrical supply to engine.

firewall it — use all available power for speed or climb. See also **cobbed it.** —from Paul Turk. In early planes, the throttle on the "dashboard" was not only on the "firewall" but also was pushed in for acceleration, rather than pulled out as it was in older cars. Aircraft often have firewalls between engine ahead and cockpit behind; thus slamming it forward **firewalls** it. See **through the gates.**

First Officer — not the captain, but the copilot, who flies in the righthand seat in the cockpit. It is said that a First Officer need know only two things: "All clear on the right" and "I'll have the chicken" (because the captain would always take the steak). This second-in-command-pilot has three stripes on the sleeve of his uniform.

first one on the ground — result of an unopened 'chute, what a parachutist does not want to be, as will be clear from a perusal of **cigarette roll.** May also happen after an occurrence of a **Mae West** (sb. 3).

fish finder — slang for onboard traffic avoidance radar.

fishheads — other military pilots' term for Navy pilots.

Fishpond — in WWII military aviation, late in the war (1944), daytime bomber sorties by the RAF, escorted by fighters. Earlier ones were called **circus.** —Doug Sample CD, Canadian Branch, Yorkshire Air Museum.

fishtailing, fishtail landing — (from early aviation) "swinging from side to side to retard the forward speed, as when landing in a small field." —Berrey and Van den

Bark, *The American Thesaurus of Slang*.

fitter — British term for an aircraft engine mechanic. See **rigger**.

five-mile club — members are pilots who have flown above 25,000 feet altitude (perhaps a take-off on the **Mile-High Club**, couples who have made love at or above 5000 feet).

fix — to establish one's position through any of several navigational aids or techniques.

fixed pitch — non-variable propeller blades (they rotate or spin, but they don't vary in their angle to the forward direction).

fixed wing — conventional wing planform as opposed to **rotary wing**, as in helicopters.

fizzer — to be put up on a charge. WWII. —Doug Sample CD, Canadian Branch, Yorkshire Air Museum. "For such a felony, Plonk can be slapped on a **fizzer**, brought up before his commanding officer, and possibly achieve a Court-Martial." —Raff and Anthony Armstrong, *Plonk's Party*.

flag — a red tag on a navigational instrument to alert pilot that the instrument is inoperative or too far from the **navaid** to which it is tuned.

flag carrier — the national air line of a country, often wholly or partially government-owned, e.g. Trans Canada Airlines, Air Canada, BOAC, Air China, Sabena, Lufthansa, Alitalia, El Al, Finnair, KLM. While many of these airlines were at one time government-owned, the trend is toward privatization: BA has not been government-owned since Thatcher.

flak — According to Partridge, from the German *Flugabwehrkanonen*, or anti-aircraft guns. German dictionaries list both this word and *Fliegerabwehrgeschütze* as terms for "anti-aircraft guns." See **Archie**.

flak-happy — trembling of hands, the **twitch**, etc. caused by exposure to **flak**. —Douglas Sample CD, Canadian Branch, Yorkshire Air Museum.

flamarino — RAF WWI term for German plane going down.

flame — as a verb, to shoot down an enemy aircraft in flames.

flame-out — engine failure.

flaming onion — a type of anti-aircraft shell which, when exploded, has the appearance of a flaming onion.

flaperons — trailing-edge control surfaces incorporating the properties of both flaps and ailerons, such as on the Boeing L-15 Scout.

flaps — technically "control surfaces," e.g. spoilerons, flight spoiler panels, etc., they are commonly known, collectively and conveniently, as flaps, even among French-Canadians, whose other choice is "hypersubstantateurs"!

2. Jocularly, birds do it, but not recommended for fixed-wing aircraft.

flare — the slight increase in pitch that occurs naturally on landing, just before touchdown, due to ground effects. End powered portion of descent on landing approach, bring nose up, let aircraft settle down onto runway. "I had to resist a tendency to flare too high," on landing (from a description of what it was like to fly a new plane, the Canadair Regional Jet), David Hughes Pilot Report in *Aviation Week and Space Technology* (Farnborough issue, September 7, 1992).

flare path — lighted runway (Brit.).

flat hat — verb, to **stunt**, show off, "grandstand." Derived from the old USAAF custom of leveling the headgear, by **buzzing**, of **groundpounder** reviewing stand officers. —Bob Stevens, *"There I was."*

flat spin — deadliest condition known to aviation resulting from failure to manoeuvre or control the aircraft adequately. If the nose is down, gravity at least will lend some speed to help **pull out**. In a flat spin, the plane becomes like a leaf, and the pilot has no **control authority**. According to Eric Hehs, "some aircraft do not recover easily from the condition. So they just drop out of the sky."

flat-top — nickname for "aircraft carrier," because of its shape.

flick roll — another name for the **snap roll**. See also **one "G" roll**.

flight — a number of **elements**. See **squadron, wing**.

flight control systems — manual, hydraulic, or electronic systems monitoring and/or operating the flight controls of an aircraft.

flight engineer — a third pilot, required in some aircraft, to "monitor the various aircraft systems during flight." Usually has three stripes on the sleeve of his uniform but may have two. — "Airline Taalk," American Airlines field training/ground services materials, March 1993.

flight level — altitude in thousands of feet, as in "flight level 24," or 24,000 feet. Usually applied to altitudes above 10,000 feet. Sometimes spelled "fl". See **angels, cherubs**.

flight-seeing — a pun on "sightseeing," denoting scenic flying, say, in Alaska, from pontoon or ski-equipped airplanes like the 6-passenger Cessna 206.

flimsy — list of aerodromes in WWII War Theatre area. Names, positions, signal letters, secret code radio call signs, which were changed every 24 hours, for use by aircraft over enemy territory. Information was printed on a lightweight piece of paper (rice paper), so that in the event of being forced down, the flimsy could be quickly destroyed, by eating it. Later applied to any piece of paper containing information passed among the flight crew. See **shrimp boat**.

flip — a ride. See **hangar rats**. 2. to make one half of a snap roll, thus flipping the airplane on its back.

flipover — Canadian military talk: a trip on a pass to spend the weekend in Europe.

flippers — slang term for elevators or horizontal control surfaces.

FLIR — Forward Looking Infra Red, a sensor device usually mounted on aircraft engaged in the ground support role.

float — the tendency of an airplane to remain airborne after the landing **flare**. See **kite**.

float plane — see **seaplane**.

flop — "make an ungraceful landing in a fighter plane, especially on an aircraft carrier.'" —Murray, "Language of Naval Fighter Pilots."

flower — code name in WWII for patrols of German airfields when the bombers stationed there were out. —McIntosh.

fluff and buff — "the new uniform standard in the AOE (American Overseas Expeditionary force) meaning wash-and-wear, fluff-dried BDUs, Battle Dress Utility, the normal work uniform used in the field, as opposed to the traditional heavily starched and pressed uniforms of the past, with creases sharp enough to shave with) and buff-polished boots (as opposed to the spit-shined, glistening footwear of the past.) —Hans Halberstadt, *Airborne: Assault from the Sky*.

fluid four — "The fluid four formation, harking back to the **thatch weave** of WWII, was used for optimum visual and radar contact, should MiGs appear. Two flights of fighters would fly one above the other, the second flight weaving above and behind the lead flight with one aircraft in each flight responsible for visual search, one for radar search." —Mersky and Polmar, *The Naval Air War in Viet Nam*.

flush launch — USAF term for a mass-launch of virtually all the aircraft on the base. See also **Alpha strike**, **MITO**.

flush rivet — headless rivet creating a smooth, low-drag skin.

fly-by — an aerial revue.

fly by light — short way to refer to automation of aircraft controls, as fiber-optic transmission of data replaces electronic means (**fly by wire**).

fly by wire — There's no direct mechanical connection between control and controlled surfaces. Electronic control. The F-16 is totally "fly by wire," and the Concorde was the first commercial airplane to be so. The computer is programmed so that fatal manoeuvres cannot be done by manual override; "the ape can't kill himself" (What must be avoided is loss of control resulting from "G-lock", "G-force induced loss of consciousness." The black (control, computer) boxes slow down the instructions and allow automatic searches for the right setting to take place in interface with the controlling instructions). In some aircraft, the plane is dynamically unstable unless the computer is available to give it multi-active input. A term adapted to car talk in the form "drive by wire." On its way to being superseded, in aircraft, by **fly by light**. One recent use suggests an ambiguity in the term or a lack of widespread precise understanding of its meaning: "All modern aircraft are automated to a degree, but design traditions in Europe and North America differ. North American manufacturers like Seattle-based Boeing Co. put the pilot at the centre of things, making the aircraft *almost* **fly by wire**: European manufacturers go all out for hi-tech, making the computer king." —Jeff Heinrich, "Popular 'John Wayne' Airbus in no danger of flying into sunset," Montréal *Gazette*, June 17, 1994. Paul Dickson has it as "to fly by autopilot".

A "fly-by-wire" pilot's joke asks, how many crew are necessary to fly one of these new aircraft? A pilot and a dog: the dog is to bite the pilot if he touches the controls, and the pilot is to feed the dog. —from the television special on the new Boeing 777, "21st Century Jet."

fly caps; fly escort; fly sweeps — "to fly on the indicated type of mission (Canadian source)" —From John and Adele Algeo, "Among the New Words: Gulf War Glossary," *American Speech* (Winter 1991).

flyers and talkers — term used in Viet Nam meaning rockets and 7.62 mm machine gun ammo. —Ray Seale.

fly-in — like a drive-in, with aircraft. Usually to an airshow or gathering of **birdpeople**.

flying — "the art of avoiding ongoing disaster." —Ted Beaudoin, *Walking on Air.*

—"Flying, like the sea, is not inherently dangerous. But, it is mercilessly unforgiving of human error." — Hank Searls, *The Crowded Sky.*

flying a desk — said of a pilot assigned to a desk job. See **caged bird**.

flying the backside of the clock — pilots' term for what they do when they fly freight, mostly at night. See **freight dogs**.

flying the beam — a hazing stunt inflicted upon student pilots during WWII wherein they had to walk along a straight line with their arms held fully out to the sides, like a set of wings. 2. More traditionally, it refers to following a navigational radio beam. See also **iron beam**.

flying boat — actually a boat that can fly, combining the aerodynamic characteristics of the airplane with the hydrodynamics of the boat. Aloft, they react to control surfaces on air; afloat, they're controlled like a boat. A good flying boat pilot is able both as a boatsman and as a flyer. The fuselage of a flying boat is often called the **hull**, especially when referring to the boat-like lower portion. See **sea-**

plane, sponson, hydrofoil, weathercock.

Flying Boot Club — its exclusive members returned on foot from German POW camps. Doug Sample recalls that the flying boot with single wing was worn by anyone who had "evaded capture and 'walked back.'" See also **Gold Fish Club, Winged Boot, Caterpillar Club, Walker's Club**.

flying the cockpit — term used to describe instrument flight, See **blind flying, on the clocks, on the gaages**.

flying coffins — early term for airplanes, according to the PBS special "The Great Air Race of 1924." Specifically applied to the German fighter-bomber JU-88, according to a Luftwaffe pilot interviewed on the television show Wings.

flying fever — mad or crazy about flying and/or aviation. See **aeromad**.

flying horse — early term for an airplane when it was still considered an adjunct to the horse cavalry, pre-WWI.

flying in tight formation — from its usual meaning, to describe a number of aircraft moving in close proximity and synchronization, the term was incorporated into a joking description of the Boeing 777, as "over three million parts flying in tight formation," in the television special on the building of the aircraft, "21st-Century Jet."

flying lawnmower — term used to describe all manner of flying machines from powered hang gliders to microlights.

flying machine — very old term for "airplane." Its antiquity is celebrated in an old RAF song: "Come Angeline in my flying machine / We'll go up, up, up."

flying rock — "If he turned back to Albuquerque and completely lost communications, he would become a 'flying rock,' a menace to the airways because he was not carrying out his original flight plan." —Hank Searls, *The Crowded Sky*.

Flying Seven — first women's flying club in Canada (the **99s** then being limited to Americans), formed in the late 1930s by seven women pilots in Vancouver. They wore a distinctive divided skirt and Glengarry hat. —Joyce Spring, *Daring Lady Flyers*.

flying speed — the speed at which the airfoil produces enough lift to unstick the undercarriage. "Just above falling speed." — Gordon Baxter, *More Bax Seats*, 1988.

flying tail — an empennage design in which the entire tail assembly is moved as part of the elevator movement.

flying the Hump — WWII expression to describe the trans-Himalayan cargo run between bases in India and forward bases in China.

Flying Tigers — The **AVG,** American Volunteer Group, sent to China in the 1930s under the command of Gen. Claire Chennault, flying P-40s. After the war, the name of a cargo airline set up by Chennault.

2. Also, a group of Russian pilots, flying Mi 26 helicopters into Bosnia on relief missions, paid in American dollars, described as "international humanitarians involved in clandestine relief drops." CBC News, November 9, 1992.

flying wing — appellation given a series of Northrop designs that were tailless, basically a pair of delta wings molded in the middle into a fuselage. Applied generically to any such design.

fly it before you buy it — "Bluesuiters," government aviation purchasing officials, "own inviolate rule," not followed in the case of the **Have Blue** Lockheed stealth fighter project development, according to Ben Rich in *Skunk Works.*

fly-off — a competition, usually aircraft competing for a government contact, but sometimes applied to any flying competition.

flypast — aircraft, often in formation, fly near a reviewing stand or crowd, usually in the presence of dignitaries.

fly streamers — "to be shot down." —Berrey and Van den Bark, *The American Thesaurus of Slang*.

2. Student pilots on their solo flights, or with very few solo hours, flew red pennants from their wing tips to warn other pilots of their novice status. WWII.

fog — "These [air] regulations were certainly all-embracing and I remember still that an aircraft in fog was to 'proceed with caution making a loud noise such as ringing a bell.'" —Group Capt. J. A. Kent, *One of the Few*.

fog buster — nickname given by Horizon Air for its **head-up-display** system for the de Havilland Dash 8-100 that will allow pilots to fly Category IIIa instrument landing system weather conditions ... down to a 50-foot decision height and a runway visual range of 700 feet.

fog factory — the region of the Bering Sea where fogs originate, according to Assen Jordanoff, *Jordanoff's Illustrated Aviation Dictionary*.

fogged in — completely covered by fog, forced to stay aloft or on the ground because the airport is fogged in.

Fokker construction — term applied to welded steel fuselage and plywood-covered cantilever wing generally employed by Fokker aircraft, designed by Reinhold Platz from 1916 to the 1930s.

Fokker fodder — said of various vulnerable, inefficient Allied warplanes of WWI, such as the British BE 2, which was often easy prey to the much superior Fokkers.

Fokker scourge — said by Allied airmen in WWI of the deadly effectiveness of the mostly Fokker-equipped German air force, before Allied pursuits became vastly improved.

forced landing — an emergency landing. "There is no such thing; all airplanes eventually come down, they do not need to be forced." — Gordon Baxter, *More Bax Seats*, 1988.

formation, flying in — technically (and especially militarily), several aircraft flying at specified altitudes and at precise distances apart, in a constant and agreed-upon design. It entered slang when "Fed[eral] Ex[press] chairman Fred Smith had been lobbying federal aviation authorities...for an amendment that would allow him to operate larger aircraft. Because of the small jets (Falcons) Smith was using, people joked that he had to fly them in formation to get all his parcels to their destinations.....″ — from Stuart Logie, *Winging It*.

forty-mile-long-gorilla — "strike package" of fifty to sixty aircraft, during the Gulf War. —U. S. News and World Report, *Triumph without Victory*, 1992.

forward fire control — see **bird-dog**.

foul deck — in naval aviation this means a flight deck which is not ready to receive or launch aircraft. Opposite of **clear deck**. See **ready deck, green deck**.

Four Corners High — US Weather Service term for a summer monsoon pattern typical to the Four Corners region where Colorado, Utah, Arizona and New Mexico meet. The stationary high pulls warm, moist unstable air from the south, resulting in scattered clouds in the morning, building to big thunderstorms in the afternoon.

four-ship — a four-fighter pack, during the Gulf War.

Fowler flap — full span trailing edge flap named for its inventor.

Foxtrot Oboe — "we foxtrotted oboe" means "we fucked off". Pilot slang of an old air force type, using the old code alphabetic words ("Able," "Baker," "Charlie," etc.) With the current code words, this would now be "Foxtrot Oscar."

free hunt — ("freijagd") German expression for free-ranging fighter sweeps over England, usually by Messerschmitt Bf-109s. —Deighton, *Fighter*.

freight dogs — pilots who fly cargo duty, often using old, dirty planes, often at night; and the planes they fly. See also **backside of the clock, flying**.

French kiss — among "stewardesses," "a midair collision," according to John Davis, *Buzzwords*. See also **bingo**.

French landing — a landing made on the two main wheels with the tail held in the flying position as long as possible.

front man — what the **GIB** (the guy in back) calls the guy in front. Usually the pilot, but on some aircraft could refer to the bombardier, the radar operator or the weapons operator, if their position is in the nose.

fuel cells — fuel tanks, often collapsible or temporary.

fuel strainer drainer container — a tall narrow vessel used during **preflight** checks to drain fuel from tanks in order to check for water or contaminants.

fruit salad — ribbons and medals worn on the breast of a tunic. See **scrambled eggs**.

full throttle landing, full throttle descent — at high speed, once reported as a manoeuvre to deceive hijackers into thinking New York's JFK airport was Havana, Cuba, by landing rapidly at a deserted corner of the field.

full tray — among "stewardesses," a term for an overcrowded airplane, according to John Davis, *Buzzwords*.

FURAT — "further information at....(time or place)" in flight attendant and ticket agent talk. —from John Cavill.

fur ball — "Confused tangle of a **dogfight** with air-to-air missile fire." —From John and Adele Algeo, "Among the New Words: Gulf War Glossary," *American Speech* (Winter 1991). Also explained as "a turning fight involving multiple aircraft" in Canadian Forces Aircom Regulation Glossary.

G, G-force — gravity, the stress on equipment and human body measurable in units of G-1, G-2, etc. "That physical phenomenum that causes your head to sink just above your beltline while rat racing a jet fighter." — Stevens, *"There I was."* See also **Bavarian test pilots**.

G-force facelift — "ten positive Gs will push gravity against your face until your ears lie behind your head." —Dawn Hanna, "Chileans display attitude at altitude," Vancouver *Sun* (August 11, 1995).

gaggle — "RAF bomber formation - lines of five aircraft abreast, each 200 yards apart and each rank 300 yards behind the one ahead. Every plane flew at a different height as well, so that, while they were a most dispersed target for flak, they could converge on the target and bomb almost together." — Brickhill, *The Dam Busters*. Doug Sample recalls that the "gaggle" was a loose formation and was a word used during the late stages of WWII.

Galland hood — an improved clear-view canopy designed for the Messerschmitt Bf-109, named after the famous Luftwaffe WWII ace, Adolph Galland.

gardening — sea mining by WWII aircraft. See also **vegetables**. —Doug Sample CD, Canadian Branch, Yorkshire Air Museum.

gas — in modern usage always fuel, but historically, that which provided the **go. Juice.**

gascolater — a fuel strainer. See **fuel strainer drainer container.**

gash aircraft — RAF, any aircraft that does not show on the official inventory.

gate guardian — a retired aircraft on permanent static display, typically near the gate of a military establishment.

gear-up-you-got-it, gear-down-I-got-it — a breed of **captain** who handles all the takeoffs and landings, but lets the **first officer** do all the enroute flying.

gen — British WWII word for "information," as received, e.g. in briefings. — Berger, *Invasions without Tears: The Story of Canada's Top-scoring Spitfire Wing in Europe during the Second World War*. Hence the title of the glossary *What's the Gen?* by H. W. about 1944. "Gen" was short for "general information."

general aviation — term used to describe the realm of private and corporate aviation, as opposed to commercial or military aviation. The **ICAO** 50th anniversary book defines it as "an aircraft operation other than a commercial air transport operation or an aerial work operation."

Gens de l'Air — Canadian French organization of airline personnel, including some pilots. Came to wide attention during the agitation for the extension of official government bilingual policies to the language used between pilots and air traffic controllers, during the Trudeau era.

George — slang for the automatic pilot. WWII. "Who's flying the airplane, me or George?" **Autopilot.**

geriatric jets — an airworthiness term used for old aircraft which have completed their finite lives.

Gib, gibs — guy in the back, "Guy in the back seat, the electronics officer in the rear of a two-seat aircraft" — From John and Adele Algeo, "Among the New Words: Gulf War Glossary," *American Speech* (Winter 1991). The elecronics officer could be a Radio Intercept Officer or an Airborne Radar Technician.

gimper — "an aviator who is known as a good fellow, a superior aviator." — Rose, *A Thesaurus of Slang*. This term, which turns up in a number of slang dictionaries from the 1930s and 40s with this meaning, is used in other military slangs and in hockey to designate someone who is injured or physically unfit.

glamour boys — fighter pilots (WWII, bomber slang). —Doug Sample CD, Canadian Branch, Yorkshire Air Museum.

glass cockpit — an up-to-date cockpit with "multifunction display" — i.e. instead of all those dials, a certain number of computer terminals with screens on which one may call up information as needed, plus maybe a few dials. The Canadair Regional Jet has six screens. —from Bob Wohl. Some modern aircraft are known as three-screen, six-screen, etc. Barry Lopez describes the Boeing 777 and the Airbus 320 as being called "synecdochically as 'glass cockpits,' ...in which the information most frequently reviewed is displayed in color overlays on video-like screens."

glasshouse — RAF military prison where in WWII a serious crime punishment was served. See also **digger, cooler, moosh.** —Doug Sample CD, Canadian Branch, Yorkshire Air Museum. "Don't land with your wheels up or you'll go to the glass-house at Curry." —Williams, *The Plan.*

glassy water technique for landing — "used for setting a seaplane down on a mirror surface, that is, a surface you cannot see. (What you see is not the surface, it is a reflection, mostly of the sky. It is totally deceiving and a deadly trap for the unwary)....You adjust your approach to the lake to come in as low as possible over some positive height reference, such as, in daylight, trees along the shore. Tonight I would look for lights along the shore. As you cross the reference point and float out over the lake, you ease the nose up into the 'step' attitude—the attitude in which the seaplane runs **on-the-step** on the surface of the water—leaving on enough power so it is sinking slowly. ....[Remember] you cannot see the surface, and not to try to see it.....Keep the directional reference straight ahead. If you 'feel' the rate of sink is too low, you take off a little power; if you 'feel' the rate of sink is too fast, you add a little power, carefully maintaining the 'step' attitude. And you wait, with what patience you can command, for the floats to touch, always a complete surprise to you. It requires more room than normal, but it works." —Vi Milstead, first woman bush pilot, and a pilot in the Air Transport Auxiliary, an **atagirl,** as quoted in Spring, *Daring Lady Flyers.*

glide bombing — the target is approached in a shallow dive; the bomb, released early, glides to target. See **dive bombing.**

glideslope — descent path guidance element of an Instrument Landing System.

gliding distance — "one half the distance from an airplane to the nearest landing area at the time of engine failure." — Stevens, *"There I was."*

glitch — jargon for trouble, problem or bug.

G-lock — G-force induced loss of consciousness. See **B-lock**.

glowworm — "an anti-sub tactical manoeuver, usually performed by Grumman Avengers....a hair-raising three-part manoeuver. It began with a steep dive to increase air speed, followed by a sharp pull-up when several rocket flares were lofted into the sky. With the rockets away, a stomach-raising pushover brought the Avenger down for an attack on the illuminated sub." — Foster, *Sea Wings*.

go — the engines and/or the fuel. That which provides the "go." See **juice**.

on the go — "going round again," i.e. taking off after an aborted landing attempt to circle and attempt another landing. Some of the last words from the cockpit of the USAir Boeing 737 which crashed in Charlotte, N.C. in 1994. (Associated Press wire story in Austin *American-Statesman*.)

go-around — a **missed approach**. See **on the go**.

go team — a group within the U.S. National Transportation Safety Board which is assigned accident duty on a weekly rotation basis. Members of the team are prepared to drop what they are doing and proceed to the scene of a major accident at a moment's notice. See **tin kickers, jigsaw mullahs**.

goatscrew — "a variation on **FUBAR**, but not quite so bad. A disorganized, embarrassing, and graceless chaos." — Halberstadt, *Airborne: Assault from the Sky*.

God is my Copilot — "On June 1 Avro experts fitted new automatic pilots in the Lancasters for the D-Day operation, and Nicky Knilans at last found out why his much-cursed 'R Roger' flew like a lump of lead. They found it needed longer elevator cables than the others, inspected it to find out why, and discovered that the elevators had been put on upside down at the factory. Knilans had been flying it for months like that and as Cheshire said, 'Only you and God, Nicky, know how you stayed up.'

'Not me, sirrrr,' Knilans said in his American drawl.....'Only God. I didn't know.'" — Brickhill, *The Dam Busters*.

godunk — one who hangs around airports, bumming rides, without offering to pay for them. **Hangar rat**.

you're **go for throttle up, Challenger** — since the NASA Challenger disaster, a customary way for irreverent mechanics to say "go ahead and start it!"

go guns — portable Vickers machine guns used in training RCAF in WWII: "You carried them with you, and then mounted them on a turret thing at the back..." —Williams, *The Plan*.

going upstairs — gaining altitude, climbing.

golden BB — "Soviet-originated Iraqi defense doctrine that if a large number of projectiles are shot into the sky, some will hit a target (derisive)." —From John and Adele Algeo, "Among the New Words: Gulf War Glossary," *American Speech* (Winter 1991). Ben Rich also refers to the "curtain of steel [over Baghdad] which...looked like three dozen Fourth of July celebrations rolled into one...but represented blind firing....They hoped for a golden BB—a lucky blind shot that would hit home." — Rich, *Skunk Works*.

Golden Eagle — RCAF slang for the paymaster and/or the aircraft in which he arrived with the money bags. "Careful! Off the circuit. Get that Golden Eagle down." —Williams, *The Plan*.

Gold Fish Club — cloth patch depicting a goldfish and worn on the tunic to indicate one had **ditched**, among WWII military aviators. See also **Caterpillar Club, Winged Boot, Walkers Club, Flying Boot Club.** —Douglas Sample CD, Canadian Branch, Yorkshire Air Museum.

gone in the oggin — "went down into the ocean," British pronunciation, West Coast. "Ditched in the drink."

gong — a medal (WWII). See also **NAAFI gong.**

gonged — awarded a medal. "He wore a string of gongs." —Doug Sample CD, Canadian Branch, Yorkshire Air Museum.

Go/No Go — instrument panel indicators describing a system's operational status. "Describes, in a broader sense, if one can fly or not, for instance, a flat tire would be a *no go* to perform a flight. Or, in an electronic system something [that] does not work. One would check item by item with test equipment, which would allow [one] to find the tested item functioning or non functioning, *go or no go*." —anon. Brit. Canadair technical writer informant. See **MEL.**

good hands — also known as **good stick**, used to refer to general motor skills ability in a pilot, reliable coordination of rudder and other controls, especially in a crosswind.

goose flares — type of runway flare used in Canada during WWII. "They were like a tea kettle with a great long neck on them and filled with kerosene and a wick, and they were the filthiest things that ever happened; they were just covered with carbon. You ascertained the wind direction from the sock and laid the flares out in a T formation, and you flew into the T. I think we laid ten flares on the main leg and four across the top,." — Williams, *The Plan*.

gossport — "speaking tube between instructor and student." —Bob Stevens, *"There I was."* Alternate spelling of **gosport**.

GPS — Global Positioning System, a satellite-controlled air traffic and navigation system, also in use for ships.

grandmothers — [Bill Lear] "once said he would sell his grandmother to save a pound of aircraft weight. Shop floor jargon substituted 'grandmothers' for 'pounds.' " — Szurovy, *Learjets*.

Grand Slam — 22,000 pound bomb, also called the **earthquake bomb**. WWII. See also **tallboy, doodlebug, blockbuster, cookie**.—Doug Sample CD, Canadian Branch, Yorkshire Air Museum.

grasshopper — see **L-Bird**.

graveyard spiral — "the descending turn a pilot slips into when vertigo panics him into believing his own senses rather than his instruments." — Searls, *The Crowded Sky*.

grayout — British naval aviation term for partial loss of vision due to sustained positive acceleration. Blurred vision caused by excessive G-forces. See **blackout, brownout, whiteout, redout**.

grease it in, grease job — to make a smooth landing. See also a **CEO-is-aboard-landing, kiss landing**.

green — said of a new aircraft in fly-away condition, with **avionics** but without final paint job or interior. It has been speculated that the term comes from the colour of a substance coating the aluminium fuselage to prevent deterioration, but it is more likely that it is related to "green beer," i.e. beer not yet ready. It may be derived from "green" as in "green fruit," i.e. "fruit not ripe yet." The coating substance in Germany was in fact grey (zinc chromate), and in one pilot's early years in aviation there he asked, "Why do you call that grey plane 'green'?" See **white-tail**.

green deck — an aircraft carrier deck prepared to retrieve landing aircraft. Also known as a **ready deck**. —Coonts, *Flight of the Intruder*. See **clear deck, foul deck**.

green run — an engine endurance test on the ground.

greenhouse — cockpit canopy. Some aircraft have been called **flying greenhouse**, because of an abundance of **perspex**.

gremlin — mythical creature blamed for mechanical, etc. failures and faults. WWII, with continuing aviation use.

This term, which has come to be applied to any mysterious mechanical disfunction in almost any machine, was first used of airplanes. According to Partridge, it means "a mischievous sprite that, haunting aircraft, deludes pilots...since the late 1920s....Gremlins could even be helpful on occasion; they were 'responsible for all unaccountable happenings—good or bad.'" See **fifinella**.

gripe sheet — a fault concerning an aircraft engine, airframe, system or other equipment, is entered on a "gripe sheet" for service. See also **up gripes** and **down gripes**.

gross weight — jocularly, a pilot weighing 350 lbs. or more.

grounded — said of aircrew or aircraft when their ability to fly has been temporarily or permanently impaired. **Scrubbed, washed out.**

ground effect — the upwash of air at the wing produced by the proximity of the ground. An airplane flying close to the ground at a given angle of attack experiences an increase of lift and a reduction of induced drag, which account for the "floating" occurring in some types of aircraft at landing. According to a recent report, "Russia is working on the Ekranoplan, a revolutionary half-plane/half-ship that will cruise at 480 km/h just three metres above the water — using the so-called 'ground effect' to coast on air compressed near the Earth's surface." —"Social Studies," Toronto *Globe and Mail* (Dec. 21, 1995).

ground flying — a discussion of flying activities or experiences by pilots on the ground. Not to be confused with **armchair flying**.

ground gripper — non-flying personnel associated with aircraft. Also **deadbeat, penguin, kiwi**.

groundhog — "a pilot who does not enjoy flying." —Berrey and Van den Bark, *The American Thesaurus of Slang*. See **kiwi, penguin, ground pounder**. A person who works with aircraft, but doesn't fly.

ground loop — in the old North American **Harvard** (T-6) trainer, also called a **Texan**, they say there were two kinds of pilots, those who had had their ground loop, and those who hadn't yet had it, according to Peter Otto, postWWII

Luftwaffe **Bavarian test pilot** and later Canadair representative in Germany. If somehow the tail wheel (the aircraft was a **taildragger**) had gotten turned sideways, at right angles to the front of the aircraft, "because the tailwheel was unlocked or the tailwheel locking device was broken, the aircraft tended (particularly in gusty weather and as a result of overcontrolling) to break out sideways and turn uncontrollably about its vertical axis, round and round, and you could not do anything but wait for the stop." This can happen to any aircraft with a **narrow track**. The Messerschmitt Bf-109 had this characteristic. So did the Stearman PT-13 Kaydet, dubbed for this reason the **Yellow Peril**. "Given a moment's inattention, a Stearman would ground loop into the middle of next week." — Bill Marsano, *Air & Space Magazine*, Oct./Nov. 1994. He defines a ground loop thusly: "If during the landing rollout the airplane begins to swerve or the brakes are injudiciously applied, the center of gravity tends to take the lead, and, if unchecked, the tail will swap places with the nose."

ground pounder — a person who works with aircraft, but doesn't fly. See **ground hog**, **kiwi** and **penguin**.

ground return — confused blips on a radar screen caused by echo from terrain. Low-flying aircraft can be lost in "ground return."

ground speed — actual speed over the ground, as affected by head, tail, and cross winds.

Groupie — RAF WWII slang for Group Captain.

G-string — type of safety strap on some aircraft in WWII. "They had a G-string that held you in so you wouldn't go into the Wild Blue." — Williams, *The Plan*. See **jockstrap**.

G-suit — "jumpsuit worn by crews of fighter planes; the suit has inflatable air pockets that are used to force blood from the lower limbs back to the heart when one has to **pull Gs**, thus preventing its wearer from blacking out." —Murray, "....Navy Fighter Pilots".

guard — the emergency frequency - 121.5 MHz; military 243.0 MHz. "I sent my **Mayday** out on **guard**." Known derisively to USAF pilots as the "Navy common" channel.

the Guinea Pigs — WWII burn victims, especially air crew, who were taken to an experimental burn surgery unit near London for treatment. The large number of victims brought about the invention of new techniques for reconstructive surgery.

gull wing — in which the wing slants sharply upward from the fuselage, then straightens out. See **inverted gull wing**.

GUMP check — Gas Undercarriage Mixture Pump. — Stevens, *"There I was."* John Race, Canadair Challenger pilot, says the "p" stands for "propeller." See **PUFF check**.

guy driving — term applied to the person actually flying the airplane at a given time.

hairy — a wild, frightening incident, as in "a hairy landing."

hairy manoeuver — "...results of not keeping forward pressure on the stick while inverted in a slow roll. The airplane will drop its nose and race for earth, quickly gaining speed." — Johnson, *Thunderbolt*.

hairy-op — a bomber sortie that encountered mechanical problems, fighter attacks, flak damage, etc. WWII. See also **shakey-do**. —Doug Sample CD, Canadian Branch, Yorkshire Air Museum.

Halibagger — a WWII pilot who flew the Handley-Page Halifax bomber, which was also known as the **Halibag**. —from Doug Sample, former **tail-end Charlie** on Halibags out of Yorkshire during WWII.

Halo jump — High Altitude Low Oxygen jump, made from an aircraft at great height. —Pat Dudley.

hammerhead stall — an aerobatic manoeuver in which the aircraft climbs almost vertically to the point of stalling, then is tipped left or right in a **wingover**, resulting in a 180-degree change of direction.

hands off — flying a perfectly trimmed aircraft, or one on **autopilot**. See **heads up, hands off**.

hangar flying — talking about flying, as distinguished from really doing it. Known also, especially when exaggeration is present or suspected, as "to **start the lamp swinging, line-shoot**.

hangar session — a meeting of aviators, regardless of whether held inside a hangar or not.

hangar pilot — the mechanic who sits in the pilot's seat and manipulates controls and brakes while others push or tow the airplane into or out of the hangar.

hangar queen — jocular insulting name for an airplane which is hard to sell, or for a craft that is constantly out of service because of excessive repairs or maintenance. For Canadair, the CL-84 has been called a hangar queen (Logie's *Winging it: The Story of Canadair's Challenger)*; however, it never went into production, was produced in only four test vehicles, and, according to a Canadair spokesperson, "was highly innovative; definitely not a lemon." A true "hangar queen" is often used to provide spare parts for other planes. Also applied to aircraft that, for any reason, spend more time in the hangar than in the air.

hangar rash — "annoying dings, scratches, and broken navigational lights which aircraft acquire when pilots squeeze six planes into a four-plane hangar." — Parkin, *Wet-Coast Words*. See also **ramp rash**.

hangar rats — "the youngsters who haunted the airport ready to swing props, pump fuel, sweep floors, run errands or peddle plane rides to the public, anything to earn a little money for flying or to cadge a free **flip** in one of the trainers." —from Keith, *Bush Pilot with a Briefcase*. **Godunk.**

hanging on the prop — a low-speed, nose-high stall landing.

hang out the laundry — dropping paratroops. WWII. —*AAF: The Official Guide to the Army Air Forces*.

Happy Valley — jocular, black-comic term for the Ruhr Valley among bomber and fighter crews in WWII, because of the enormous number of anti-aircraft guns there. See **flak-happy, Ruhrflacken.**

—Douglas Sample CD, Canadian Branch, Yorkshire Air Museum.

hard point — strengthened point beneath the wing onto which external loads such as bombs or **drop tanks** may be attached. See **wet point** and **hot point.**

hard time — actual time elapsed, disregarding time zone changes. "Any route seems **long and thin** if you're flying a half-empty plane at midnight in hard time."

hard wing — one without leading-edge slats.

harmonize — what an armourer (WWII Brit. and Canadian) did to set the guns in the wing of a fighter or the turret of a bomber to ensure the bullets would concentrate in a cone at a given range. —Doug Sample.

HCU — heavy conversion unit (WWII Brit. and Canadian). Used when bomber crews switched from, typically, the twin-engine Wellington aircraft to four-engine bombers and added a seventh crew member, a flight engineer.

have the... — report of visual contact, usually with the airport or other traffic, as in "we have the airport."

have the numbers — expression used to inform **ATC** that one has the current weather, wind, active runway, etc.

hawk — passenger who causes trouble on a commercial flight. See also **vulture**. — Dickson, *Slang!*

headache — see **dog-leg, Aspirin**.

heads up, hands off — a description of what to do in **fly by wire**.

heads down — a play on "heads up," from **HUD** — **head up display**; "heads down," according to Barry Lopez, describes "some copilots [who] had only hazy notions of the geography they flew over. They were inclined to fly 'heads down,' studying the route map, reviewing the flight plan (a sequence of way points, an expected fuel burn, the speed and direction of winds aloft) and watching their instruments and screens." And the copilots most often, he says, "are ...caught up in the protracted task of programming the plane's computers. ('I don't fly anymore,' they joke, 'but I can type sixty words a m-inute.')"

heat — "attention from an enemy plane, especially in the form of shooting." —Murray, "Language of Naval Fighter Pilots." See also **in heat**.

heaters only — In USAF F-16 simulated combat, "guns and AIM-9 Sidewinder missiles only." —Lans Stout, "Cross Training South Carolina Style," *Code One* magazine (October 1993).

heavy — frequently heard on the radio attached to the call sign of a flight, it means "weighing over 300,000 lbs." "On final, heavy."

heavy iron — large aircraft, bomber or transport. In general aviation, Gulfstream, Challenger, and Falcon business jets.

heavy metal — same as **heavy iron**. Pilots are called "heavy metal drivers."

hedge hop — "short flight." — Hamann, *Air Words*. Also, low flying. See also **puddle jump**.

hedge-hopping — flying so close to the ground that the pilot must climb the ship over trees and hedges. See **balls, contour flying, nap-of-the-earth**.

height — "height is like money, it is conceivable that too much might be an embarrassment, but too little is always fatal." — Lamb, *War in a Stringbag*.

helideck — a helicopter landing area, often atop a building, on a ship or an oil rig: "Jutting out over the rig's bow seven stories above the water, the 80-foot-square helideck is rising and falling 15 feet every 10 seconds or so." —*Air & Space Magazine*, March, 1996. Also known as a **helipad**.

heli-hiking/skiing — "nouveau recreations using airlifts to take enthusiasts to summits for a day of downhill skiing or alpine walking." — Parkin, *WetCoast Words*. Popular pastime in Rockies and Bugaboos. Chopper deposits skiiers on top of otherwise inaccessible mountain crest for a once-in-a-lifetime descent.

heli-logging — "the technique was first used in northwest states, but coastal B. C. is the only place in Canada where this expensive method of log removal is profitable." — Parkin, *WetCoast Words*.

helipad — an area of land, water, rooftop or other structure used for the landing and takeoff of helicopters, usually marked with a large circle - H.

helitorch — see **drip torch**.

hell brew — cynical nickname for aviation gasoline during WWI when being shot down often meant falling to a fiery death because of it. Also called **infernal liquid, orange death, witches' water**. "Elijah was reputed to be the patron saint of aviators, but as he went to Heaven in a chariot of fire, this was something we weren't too keen about." — Kiffin Rockwell, WWI pilot

hell hole — small compartment in aft of airplane, often full of electronics and hard to work in. —Gordon Gilbert, editor, *Business and Commercial Aviation*, and a pilot. Also reported in use among U. S. Navy fliers during the 1950s.

hemp stretcher — member of balloon outfit. WWII. —*AAF: The Official Guide to the Army Air Forces*.

Hershey bar wings — squared, continuous chord, boxy wings typical to a number of Piper designs.

he's off the localizer and missing the approach — the guy's nuts.

hick hop — "carrying a few passengers a few times around the field for small sums of money." — Hamann, *Air Words*. See **barnstorming**, **barf hop**.

high Alpha — high angle of attack. See **attitude**.

high and tight — "the most popular and trendy hairstyle among the troops [of the American Overseas Expeditionary force]; not quite shaved on the sides, with about a half-inch on top." — Halberstadt, *Airborne: Assault from the Sky*.

high-hat — "a **stunt** flyer." — Rose, *A Thesaurus of Slang*. Also, "an expert stunt flyer," from Paul Beath, "Aviation Lingo," *American Speech* (April, 1930).

high toss — "low-altitude bombing technique." —"Among the New Words," Gulf War Words supplement, *American Speech*, 1992. See also **over the shoulder toss**.

high tow — the best position for a glider under tow, above the slipstream of the tow plane. See **low tow**.

hit an open switch — follow the wrong branch of a railway when using the **iron beam** for navigation. See **iron compass, Bradshawing**.

hitchhiker — "someone without a ticket," among "stewardesses," according to John Davis, *Buzzwords*.

hit the deck — literally, to land, but sometimes applied to low flying, **on the deck**.

hit the silk — bail out, with parachute (made of silk, originally, now ripstop nylon). See **silk canopy**.

hit the tanker — line up for air-to-air refueling, not, as one might think, to collide with the refueling tanker aircraft.

hittles — "'Kelly, the reason they call them missiles, instead of hittles, is that they miss much more than they hit.'" — Rich, *Skunk Works*.

hold — "In aviation parlance, **hold** always means to stop what you are now doing and thus to go around in a landing situation; but in everyday English it can also mean to continue what you are now doing

and thus to land in such a situation." —Cushing, *Fatal Words*, p. 11.

holding hands — "aircraft in visual formation." —from Canadian Forces Aircom Regulation Glossary. Occasionally said of crews crammed into tight compartments.

holding short — "pilots holding short for the same runway." —David Hughes, Pilot Report in *Aviation Week and Space Technology* (Farnborough issue, September 7, 1992). A typical **ATC** clearance might read, "taxi to runway 2-4 and hold short." The distinction is between "holding," which may be done on the runway, in position for takeoff, and "holding short," which means staying behind the yellow lines, keeping your aircraft clear of the runway, until you are cleared to proceed onto or across it.

Hollywood jump — a parachute jump made with "a Hershey bar and a .45," according to Pat Dudley, ex-Special Forces (US Army) paratrooper; more conventionally, "insertions" were made carrying full combat gear, known as a "combat jump."

homing — flying toward a **navaid**.

home and dry — naval aviation expression for a successful carrier landing.

home plate — "home airfield." —from a Canadian Forces Aircom Regulation Glossary.

honey wagon — the truck that services toilets on aircraft.

honk around — fly aimlessly. See **boring holes, swanning.**

hood — covered windows, canopies, etc. so the student IFR (Instrument Flight Rules) pilot has visual contact only with the instruments. See **blind flying.**

hoodwork — practicing instrument flying under the **hood.**

hook — see **arresting gear.**

hop — "combat flight." — Murray, "..Navy Fighter Pilots." In a broader sense, a flight.

hop the twig — Canadian phrase for "crash fatally." — Partridge, ...*R.A.F. Slang.*

hop out — "crash." — Murray, "..Navy Fighter Pilots."

horrido! — German fighter pilots' victory cry, WWII.

horizontally opposed engines — the saying goes, "If God had intended for man to fly with horizontally opposed engines, Pratt and Whitney would have built them that way." They are famous for radials. Horizontally opposed engines are more streamlined, but in some ways not as efficient, harder to cool (radials are easy to cool with air). However, Lycoming, Allison, de Havilland, Daimler-Benz—all horizontally-opposed.

horn — an audible warning device that alerts the pilot that he is approaching stall speed and/or that he is landing with the wheels retracted. An old story goes: a pilot was about to land with his wheels still up. The tower noticed it and repeatedly tried to warn him by radio, but he did not respond and executed a belly landing. Later, the tower operator asked the pilot why he did not respond to the warning. "Didn't you hear me calling you?" the controller inquired. "Hell," replied the pilot, "that damn horn was making so much noise I couldn't hear a thing."

hot and high — a description of performance capabilities of aircraft which can take off reliably and efficiently in high temperatures and altitudes.

HOTAS — hands-on-throttle(s)- and stick, describes the location of essential fighter controls to minimize hand movement during combat.

hot landing — a downwind landing. —Berrey and Van den Bark, *The American Thesaurus of Slang.* So called because it would always be at a higher than normal landing speed, and control would be more difficult, the aircraft tending to "kite" or **float**.

hot landing zone — under enemy fire.

hot point — any point, usually beneath the wing or on the fuselage that is electrically wired so that external loads can be plugged into the aircraft electrical system. Also, any point at which an external power source can be plugged into an aircraft. See **hard point** and **wet point**.

hot pickle — armed bomb release switch. "'Your pickle is hot,' he told the pilot, refering to the red button on the stick grip which the pilot could press to release the weapons." —Coonts, *Flight of the Intruder*. See also **pickle**.

hot rocks — a WWII and immediately after slang term for a show-off ace pilot, approximately the same as a **shit-hot** or a **top gun**. — Searls, *The Crowded Sky*.

hot wall — a heated crew or passenger cabin enclosed by a second, heated wall inside the fuselage. It's cold at altitude.

house — among "stewardesses," a fat passenger, according to John Davis, *Buzzwords*. "Watch that **closet hopper** try and get the **house** into the seat."

how do you hear me — radio phraseology for "did you **copy**?" Also "how do you read?" and "how do you copy?" Is the message clear?

the **howzitgoing curve** — the calculation of fuel burn expected from dead reckoning in relation to actual fuel burn and location (affected by winds and ground speeds) to measure progress against available fuel.

hub-and-spoke — Airline route systems feed traffic into major hubs, where connections are made, **PAX** changing planes and travelling to destination on spoke routes.

HUD — head-up display. An innovation in cockpit design by which instrument readings are projected onto the windshield (canopy) so that pilots can look at the surround as they read the instruments; however, some distraction and confusion may also be produced. See **final approach**. Confirmed by John and Adele Algeo, "Among the New Words: Gulf War Glossary," *American Speech* (Winter 1991). See **heads up, heads down**.

huff-duff — British term for HF/DF (high frequency direction finding), applied to early radar in WWII. —Deighton, *Fighter*.

huffer cart — "small vehicle used to blow air into the engines of a fighter plane to get them started." —Murray, "Language of Naval

Fighter Pilots." He comments, this phrase "personifies a machine by suggesting that it huffs (and perhaps puffs?) air into airplane engines."

hull — name applied to the **fuselage** of a **flying boat** or **seaplane**, but also applicable to any fuselage.

hunt — the tendency of an airplane to pitch wildly. See **phugoid oscillations**.

hushkit — modifications to engines, nacelles, etc., to allow older, noisier jets such as the DC-9, Boeing 707 and 727, BAC 1-11, etc. to meet increasingly stringent Stage 2 and Stage 3 noise restrictions.

hydroaeroplane — general name given to seaplanes and flying boats in the early 1900s. Glen Curtiss made the first water take-off in a hydroaeroplane in 1911.

hydroped — "Designed to be attached to the hull of flying boats, the 'Hydroped' consists of three stalky legs to which are mounted a series of hydrofoils and a water propeller. The idea was that for take-off the pilot would start the engine and engage the water propeller only. As the aircraft moved forward and began to gain speed, the 'Hydroped' would begin to lift the hull clear of the water by 'stepping up' from one hyrofoil to the next. When he reached high sped, and the bottom step, the pilot was supposed to engage the flying propeller and take off. Invented by Lt. Denniston Burney, 1912, the device proved ultimately unsuccessful, though contributed to aeronautical knowledge." — Allward, *An Illustrated History of Seaplanes and Flying Boats*.

IATA — International Air Transport Association, "interairline body to fix rates and ensure cooperation on safety procedures," formed in 1945, suceeding the International Air Traffic Association (1919-). —*Chronicle of Aviation*.

ICAO — International Civil Aviation Organization, created in 1944 "dedicated to the peaceful development of international civil aviation...., a permanent organizational and legal framework." —*Chronicle of Aviation*.

Icarus — the first flyer, from Greek mythology; son of Daedalus, who thought of attaching wings to the lad, whereupon the lad flew too high; sun melted wax and wings fell off. Occasionally a bold pilot will be said to "think he's Icarus."

icing — the formation of ice on wings, increasing weight but more disastrously, changing the shape of the airfoil, decreasing lift, and causing high risk of crash on takeoff. One whimsical rule is "a member of the crew must put down a hot cup of coffee on the wing surface to check" for this condition. Notoriously hard to see, and quick to form—while waiting in line, e.g. *after* de-icing. Adage: "there's no such thing as 'a little ice.'" See **cold-soaked fuel**.

ident — radio lingo meaning "push the ident button on the transponder." See **squawk**.

I. F. F. — Identification, Friend or Foe. Call for transmission of a blip to British radar screens, to identify one as Friendly aircraft. Today used to mean transponder output. In U.S. usage during WWII it meant transmission of a specific password to friendly ground stations.

IFR — Instrument Flight Rules. Some pilots say they 'fly IFR, but not hard IFR'. According to J. Mac McClellan, writing in *Flying Magazine*, Oct. 1994, "There is only one kind of IFR flying, and it's hard if you don't know how to do it." Generally, the term applies to flying without maintaining visual ground contact, on the gauges. See **VFR, blind flying, fly the cockpit**.

ILS — Instrument Landing System.

Immelman — an aerobatic manoeuver, doing a half-loop, then while upside down, rolling so that the aircraft is once more flying right side up, used to gain altitude while reversing the direction of flight. Named for Max Immelman, 1890-1916, German WWI pilot. "A sharp reversal to the opposite direction." — Rose, *A Thesaurus of Slang*.

"I do not employ tricks when I attack," he said. Indeed, nowhere in his letters or other writings did Immelmann mention using the manoeuvre with which his name is connected to this day: the Immelman turn, a half-loop with a half roll on top. One British source suggests the manoeuvre was not Immelmann's at all, but got its name from an Allied pilot who used it to escape from a tail-end attack by the feared German." —Bowen, *Knights of the Air*.

indians — small private aircraft, especially in congested airspace. — Dickson, *Slang!* The term probably derives from the Piper custom of naming their airplanes, such as Cherokee, Comanche, Apache, etc. See also **bug smashers**.

infernal liquid — cynical nickname for aviation gasoline during WWI when being shot down often meant falling to a fiery death because of it. Also called **hell brew, orange death, witches water**.

in heat — "being pursued or shot at by enemy planes." —Murray, "Language of Naval Fighter Pilots."

in-line engine — all cylinders in a row, either straight, horizontal, opposed, or vee.

interline — airline language for PAX changing from one carrier to another, or "interlining."

interrupter gear — German WWI invention which synchronized machine guns to propeller spin, allowing guns to fire through prop. Far superior to Allies' method of sheathing prop with deflectors so bullets would glance off, which was not always effective. See **synchronizer**.

intersection — the point where two navigational radio signals intersect.

in the slot — on glide slope and localizer (cone-shaped directional beam), ready to land.

inverted gull wing — a wing which slants sharply downward from the fuselage, then straightens out, such as on the Vought F-4U Corsair or the Junkers Ju-87 Stuka. See **gull-wing**.

I R A N — Inspect and Repair as Necessary. (Military slang).

iron beam — see **iron compass**.

iron compass — "the pilot's friend, the railroad," as a guide to navigation, in early bush pilot days. — Keith, *Bush Pilot with a Briefcase: The happy-go-lucky story of Grant McConachie*. Also known as the **iron beam**. See also **Bradshawing**. Using the railroad for directional indication was so general that in the 1940s, "to follow the wrong branch of a railroad" was to **hit an open switch**. —Berrey and Van den Bark, *The American Thesaurus of Slang*. See also **flying the beam**.

Irving suit — type of flying suit worn by RAF pilots in WWII. —Deighton, *Fighter*.

island — the superstructure above the flight deck of an aircraft carrier.

jam — when guns fail to continue firing because of an obstruction, usually caused by a problem in the metal links in the ammo belt.

2. to interfere with hostile radar or radio transmissions.

jammer — "Airplane equipped with jamming transmitters that produce electrical noise and false echoes to create the images of false targets on a radar screen." —From John and Adele Algeo, "Among the New Words: Gulf War Glossary," *American Speech* (Winter 1991). **Chaff** was a mechanical device to produce radar confusion; a "jammer" is electrical.

JAARS — The Jungle Aircraft and Radio Service, a missionary organization operating worldwide, flying mostly Helio Couriers to remote areas.

Jabo — *Jagdbomber*, German WWII term for a raid by Luftwaffe fighter-bombers.

JATO — jet-assisted takeoff. See also **RATO, RATOG.**

jazz music — see **schräge musik.**

J.C. manoeuvre — overcontrolling, resulting in **porpoising.** So called because after loss of control, the last thing the tower hears is "Jesus Christ!"

Jesus nut — on top of the **Huey** helicopter, the nut, with cotter pin, which must remain in place to keep the rotors on the aircraft and thus ensure that it does not fall out of the sky. The first thing the pilot says after he discovers it has worked loose is "Oh, Jesus!"

let's jet — latest way (1994) to say "let's go," in New York **jet set** talk..

jet-job — WWII term for the new jet aircraft, which Germany had and the Allies didn't. — Berger, *Invasions Without Tears.*

jet-jock, jet-jockey — "Jet plane flier." —From John and Adele Algeo, "Among the New Words: Gulf War Glossary," *American Speech* (Winter 1991).

jet lag — disrupted circadian rhythm.

jet off take — from the Chuck Berry song "No Money Down." A bit of airplane slang in Black talk. See **power brake.**

jet prop — aircraft whose propellers are driven by a jet (turbine) engine. Also called **turboprop.**

jet set — contemporary slang, in general use, "yuppie," derived from the spread of air travel and the facilitation of air travel through the creation of the well-heeled, idle, stylish youth economy.

jetstream — "When the introduction of the jets brought hopes of 'flying above the weather', jetstream became 'the dirtiest word' in the air. This high-velocity wind, circling the earth along a narrow path some 30,000 feet up, offered extra speed and fuel saving free rides to aircraft, but it was soon to be classified as an associate of murderous downdraughts". — Barlay, *The Final Call.*

jetway — a covered walkway from terminal to aircraft for the loading of passengers. Also **airbridge,** and, in American Airlines training materials, "jet bridge."

jigsaw mullahs — "aircraft accident investigators." Also known as **tin kickers**. See **go team**.

jink — "flip the steering mechanism of a fighter plane from side to side, causing the plane to roll back and forth and enabling the crew to spot any **bogeys** that may be approaching from the rear." — Murray, "..Navy Fighter Pilots." "Weaving and twisting to avoid ground fire" — Chuck Yeager.

jock, jockey — pilot, often paired with plane-type, such as **Sabre-jock**. See also **puke**.

jockstrap — safety wire in a Fairey Swordfish which clipped on to a harness and ran between the legs. See **G-string**.

Joker — "fuel state above **Bingo** for one engagement." —Canadian Forces Aircom Regulation Glossary.

joy stick — the control lever, normally extending upwards between the pilot's legs, from the cockpit floor. Used to control elevators and ailerons. In some late model jets, to one side, then called a **side stick**. The Airbus A-340 has two small ones, one on each side. The new Boeing 777 uses the old system, with a big "control yoke" mid-cockpit. **Control column, control yoke, pole, stick, pipe, yoke**. For the meaning of **joy-stick landing**, see **dead stick landing**.

Judy — "Aircrew has radar/visual contact on the correct target, has taken control of the intercept and only requires situation awareness information.....minimize radio transmissions." —Canadian Forces Aircom Regulation Glossary.

jugful — "a radio order to hold a plane." — Weseen, *A Dictionary of American Slang*.

jugs — slang for piston cylinders in an engine.

juice — slang for fuel, gasoline.

jumbo — as in "jumbo jet," refers to the largest passenger transport aircraft, e.g. Boeing 747; according to John Wheeler, Boeing public relations, first applied by the Japanese, then spread to the rest of the global aviation speech community. **Wide body, whale**.

jump jet — generically applied to a vertical take-off/landing jet aircraft, such as the Harrier.

jumpsack — parachute. "That's all, brother / Hit the jumpsack, / Bid farewell to your 39." (Bell P-39 Airacobra). — Stevens, *"There I Was...."*. Also **'chute, silk canopy**.

jump-seat sniffer — derogatory "stew talk" for a passenger who chases stewardesses, who often sit in the plane's jump seats ("retractable seat Flight Attendants occupy for take-off and landing" —American Airlines training materials). Term first given wide circulation in Jay David's *Sex and the Single Stewardess*. See also **lobby locust**. — Dickson, *Slang!*

junior jets — new term for new aircraft, such as the Bombardier Canadair Regional Jet, 50-seats, introduced in 1993, and the Empresa Brasileira Embraer-145, also a 50 seater. According to David Cay Johnston in the New York *Times* in August, 1995, the 50-80 seat jets are replacing turboprops in "a new wave that might be called junior jets."

jury rudder — because some aircraft, typically the Fairey Swordfish, produce heavy torque, pilots experimented with 'sticking plaster and gum' down the right side of the rudder to relieve pressure on the right rudder bar or pedal. — Lamb, *War in a Stringbag*. See also **clooge**.

kamikaze — Japanese word meaning "divine wind," applied to suicide missions wherein a pilot would dive his bomb-laden aircraft into an enemy target, usually a ship. Word has since been generally applied to any hazardous or suicidal mission or aircraft.

Kangaroo route — Europe - Australia air route.

Kansas — "any good flying country—open flat fields." —Berrey and Van den Bark, *The American Thesaurus of Slang*.

keep 'em flying — slogan used during WWII to encourage the sale of War Bonds. See also **plumber**.

kerosene burner — an aircraft with a jet or turboprop engine, or such an engine.

kewie bird — "an aviator who flies little but talks much about his prowess." — Weseen, *A Dictionary of American Slang*. Also spelled "kiwi bird."

Keys — (KEAS) high-tech word for knots equivalent airspeed, essentially, **indicated airspeed**.

get the **killer item** — airline pilots' oral rule. "Fog is not a killer item; a mountain is. Before you fly, know what the killer item is." —David Hoover.

kill removal — "a pilot getting clear of the fight after he's been shot down." —Lans Stout, "Cross Training South Carolina Style," *Code One* magazine (October 1993).

kipper kite — "coastal command aircraft which convoys fishing boats." —W. H. *What's the Gen?* Doug Sample points out that they protected the North Sea fishing fleet during WWII.

kiss landing — a perfect touchdown. Also known as a **CEO-is-aboard-landing, greasing it in, grease job.** — *Flug Revue* # 2, June 1992.

kiss-your-ass-goodbye position — for crash landing, the brace position illustrated in the emergency procedures card. Jocularly, "Remove glasses, remove sharp objects from pockets, put your head between your legs, and kiss your ass goodbye!"

kite — airplane. —W. H. *What's the Gen?* For another meaning of "kite," see **float, hot landing.**

kitplane — any airplane purchased from the manufacturer as a kit to be assembled by the buyer.

kiwi — student pilot, WWII. — Dickson, *Slang!* According to tradition, the "kiwi" pilot must wear his goggles below his chin while on the ground, and his heavy fleece-lined jacket zippered all the way to the throat.

2. "A kiwi is an Australian [sic: New Zealand] bird which cannot fly. The term as applied in aviation is a pilot who does not like to fly and never flies except when he is forced to do so. Same as **ground hog.**" —Dugal, *Readings on Avia-*tion (1929). **Ground pounder, kewie bird.**

KLM — "Kill 'em," joke explanation for the name of this fine Dutch airline. The name actually came from Koninglijk Lucht Maatschappij (Royal Air Company), better known as Royal Dutch Airlines.

knockout leader — air assault number one, according to *Contact*, newsletter of the Commonwealth Air Training Plan Museum, Oct. 1994.

kriegie — WWII Allied airman that had become a prisoner of war. From the German word for "war" (krieg). —Doug Sample CD, Canadian Branch, Yorkshire Air Museum.

L.A.C. — Leading Aircraftsman. See **erks.**

ladies, legal, lights, liquids — before landing, a shorthand slang way pilots on one airline reminded "crew members at 10,000 feet to turn on the seat belt sign, reduce airspeed to less than 250 knots, turn on the lights for recognition, and make sure the hydraulic pumps and fuel boost pumps are turned on." —Cushing, *Fatal Words*, p. 31.

land-on — British naval term for carrier landing. See **recovery.**

lame duck — a crippled, damaged or unserviceable aircraft. **Duff bird**.

Larry — common "stewardess" name for the copilot, according to John Davis, *Buzzwords*. See also **Moe, Curly**.

latched gate — a latched stop on the throttle preventing inadvertent movement into **beta range**, or reverse-pitch mode.

Law of Untamed Consequences — a great unrecognized rule of modern technology. In the world of cars, one of its examples is that the elimination of lead from gasoline, to combat pollution, makes the engine burn cooler; the hotter temperatures at which leaded fuel burned tended to combust away some oil deposits; thus the "cleaner" fuel is not always problem-free. Technology is full of examples of unintended side effects from "improvements," including the Dalkon shield and the Titanic, but many others came before them and no doubt others will follow. The classic moment of discovery of the Law of Untamed Consequences is contained in two lines of dialogue, unforgettable to anyone who heard them live: "You're go for throttle up, Challenger." "Uh-oh!"

lay eggs — sow mines in enemy waters, drop bombs. WWII. See also **gardening, vegetables**. —Doug Sample CD, Canadian Branch, Yorkshire Air Museum. However, according to Jordanoff, *Jordanoff's Illustrated Aviation Dictionary*, laying the eggs means dropping the bombs.

layover — aircrew term for staying overnight, resuming flight the next morning.

lazy eight — a flight manoeuver in which the nose of the aircraft is made to describe a pattern resembling a figure eight laying flat on the horizon. "Slow aileron heading changes from left to right, changing direction 225 degrees at a time." Williams, *Casey Jones Cyclopedia of Aviation Terms*, describes a lazy eight as being vertical. See **eight**.

L-Bird — slang term for "liaison aircraft," such as Piper Cub, Stinson Vigilant, for such work as observation and directing artillery at the front line. WWII, U.S. Army Air Corps. —Hardy Cannon and Bill Stratton, *Box Seat Over Hell*, excerpted in "In the beginning: The L-Birds," *Military* (April, 1992). **Grasshopper**.

lead sled — slow, heavy short-range aircraft, such as an early Lockheed L-1011. Also applied to a glider with a poor glide ratio.

Or (Brit.) the Republic F-105 **Thud**. —from Paul Turk.

leapfrog — two flight attendants working the same aisle one behind the other.

leather or feather — "The choice between filet mignon and chicken cordon bleu that pilots are offered on board" according to *Newsweek*, July 3, 1989. — Dickson, *Slang!* According to one story circulating among Air Canada pilots, the co-pilot only needs to know two things: "All clear on the right," and "I'll have the chicken," because, Captain Pete says, "the pilot always takes the steak."

left seat — flying the left seat is an expression used for the **pilot-in-command**, whether or not there is more than one seat on the flight deck. Likewise, flying the right seat is reserved for the co-pilot.

leg — 1) "non-parachute-troop soldier (paratrooper slang)" — From John and Adele Algeo, "Among the New Words: Gulf War Glossary," *American Speech* (Winter 1991). More generally used for anyone around aviators who does not fly. — Halberstadt, *Airborne: Assault from the Sky*. 2) one portion of a flight plan or manoeuvre.

leg spreader — RAF WWII term for the gold wings worn by aircrew, to which women were often sexually attracted.

let down — descent, decrease in altitude, usually preparatory to landing. See also **set down**. There's a story in the biography of Grant MacConachie about how he once explained to a passenger in bush plane days how he had decided to begin his "let down" when he did, descending miraculously through the clouds in foothills of the Rockies to arrive precisely over the Calgary Airport: he'd been smoking a cigar, and he said that when he got down to the stub, it was time to let down.

In another story, on a flight to set a new transcontinental speed record in 1945 in the Boeing XC-97. "Let-down for the landing at Washington was started over Columbus, Ohio, some 300 miles out." — Cleveland, *Boeing Trivia*.

level-off — expression used to define the process of choosing an altitude and trimming to maintain it, such as "we'll level-off at 12,000 feet."

lift dumper — flat plane flap, usually of wide span and short chord, raised from wing upper surface soon after landing to destroy lift.

lift-off — usually applied to rockets, but also said of aircraft, when they leave the ground. See **unstuck, wheels in the well**.

lighter fluid — "Doolittle put the muscle on Shell to develop a special low-vapor kerosene for high altitudes. The fuel was designated LF-1A. The rumor about the LF abbreviation was that it stood for 'lighter fluid.' The stuff smelled like lighter fluid, but a match wouldn't light it. Actually it was very similar in chemistry to a popular insecticide and bug spray of that era known as Flit. Once our airplane [the U-2 spy plane] became operational, Shell diverted tens of thousands of gallons of Flit to make LF-1A in the summer of 1955, triggering a nationwide shortage of bug spray." — Rich, *Skunk Works*.

light the wick — start (or especially re-start) the engine (especially of jets). Also **light off, lit off.** "When he had sped his turbines to 10 per cent r.p.m., he lit off with a solid thump and listened as the whistle of his jet drowned that of the starting unit." — Searls, *The Crowded Sky*.

Lilienthal — Otto, first German flyer.

Lindy — first of all a man, Charles Lindburgh, the pilot who first crossed the Atlantic solo by air. "Lindying" was an early synonym for "flying" —Berrey and Van den Bark, *The American Thesaurus of Slang*. Then, a dance, politics, kidnapping, and the establishing and ongoing presenting of the Paris Air Show, every two years, at the field, Le Bourget, where Charles Lindbergh landed.

line abreast — see **line astern.**

line astern — a borrowed nautical expression meaning one behind the other. "Line abreast" means side-by-side.

line bombing — same as **stick bombing**.

line-shoot — to boast. "To elaborate the tale of a personal experience." —W. H. *What's the Gen?* "I have long recognized the common **line-shoot** of the chaps flying certain types, generally just after their squadron has been re-equipped: 'Oh, it really is a lovely aeroplane and has a fantastic performance, but you've got to watch it -the stall is really vicious and it lands really hot. As for spinning, you've got to be right on your toes or she'll go flat on you and then all you can do is get out!' What they are really saying is: 'This is a very difficult and dangerous aeroplane. Of course I have no trouble with it, but *you* would have a hell of a time!' Of course, the aeroplane next makes its appearance as a trainer....." — Kent, *One of the Few*.

"The term originated in the days when the Service was the Royal

Navy Air Service and lines had to be shot from seaplanes to pontoons for securing." — Hooper, *Pilot Officer Prune's Picture Parade*.

listening watch — "pilots are required to maintain a listening watch on the appropriate radio frequency. In this context, it refers to the 'party line' aspect of the listening watch: pilots may overhear transmissions to other aircraft, and detect controller or pilot errors in those transmissions." —Borins, *The Language of the Skies*.

little friends — See **bomber's little friends**. Also **big friends**.

livery — aircraft manufacturers' term for the design, logo, and colour scheme to be applied to the aircraft for a particular customer, also called **colours**. See also **whitetail**.

L. M. F. — lack of moral fiber. WWII bomber acronym to indicate that someone lost courage or nerve to go on **ops**. —Doug Sample. Catchphrase applied to any aircrew member succumbing to overwhelming **twitch** and a ticket to transfer, demotion, humiliation, and often, the **digger**.

loadeo — weapons loading competitive events, in which teams compete from U.S. Air Force units, Air Force Reserve, and Air National Guard. In 1993, the fall event took place at Las Vegas and was combined with maintenance, bombing, and fighter (missile firing) competitions. —Eric Hehs, "Gunsmoke 1993," *Code One* magazine, January 1994.

lobby locust — female flight attendants' term for men who hang around hotel lobbies trying to pick them up. See also **jump-seat sniffer**. — Dickson, *Slang!*

loiter — "cruise at high speed (said of a fighter plane)." —Murray, "Language of Naval Fighter Pilots." He also comments, "'Loiter' displays irony at its best, for while its denotative meaning is 'linger aimlessly; dawdle; proceed slowly', a loitering fighter plane, as it moves through the sky at twice the speed of sound, does these things only by comparison with relatively more taxing manoeuvres such as **yanking and banking**. One cannot help believing that the person who coined it intended it as tongue-in-cheek understatement." Eric Hehs of *Code One* magazine comments, "Loiter may be ironic as applied to jet aircraft, but the word does not necessarily imply a high speed. Aircraft usually loiter to conserve fuel and stay in the same general area for a period of time."

lomcevak — a type of aerobatic manoeuvre consisting of a vertical snap and a **tailslide**. The aircraft is made to tumble tail-over-prop through one-and-a-half revolutions. According to Paul Garrison in the *Illustrated Encyclopedia of General Aviation*, "Only a handful of pilots have performed the true lomcevak, and not all aerobatic aircraft have the capability."

long hours of boredom punctuated by moments of sheer terror — cynical description of flying (and sailing).

long legs — said of an aircraft with lots of fuel, a long range aircraft — it has "long legs."

long thin route — a scheduled airline route which is relatively great in distance and has few passengers. The Canadair Regional Jet was rolled out in 1991 as an ideal aircraft for such routes, because of its size (50 passengers), fuel capacity and economy.

look down, shoot down — radar weapon system which allows aircraft overhead to monitor hostile aircraft attempting to take off. "As soon as your wheels leave the ground, you're dead." "Using look-down, shoot-down air-to-air missiles, our high-flying interceptor could defend all of North America against any long-range bomber force that would be expected to fly low to the deck to avoid radar detection." — Rich, *Skunk Works*.

look-down window — small window below the windshield for downward visibility. See also **eyebrow window**.

looker — slang term for the observer (WWI).

loop — flight in a circular path either above or below the original flight level; a vertical climb pulling back until inverted, followed by a vertical dive to level. Variations include: normal loop - a loop executed by a plane in normal level flight, beginning with a climb and ending in continued normal flight at the original level; inverted normal loop - a loop executed by a plane in inverted flight, beginning with a dive and ending in continued inverted flight at the original level; outside loop - a loop executed from normal level flight, beginning with a dive during which the plane is inverted so that the pilot is on the outside of the loop, and returned to normal flight at the top of the loop; inverted outside loop - an outside loop executed from inverted flight beginning with a climb during which a plane rolls into normal flight attitude and returns to inverted flight during the subsequent dive. —Williams, *Casey Jones Cyclopedia of Aviation Terms*.

looped — see **back to back**. "Back-to-back or looped ticketing is used mostly by business travelers to cut costs." —John and Adele Algeo, "Among the New Words," *American Speech*, Spring 1994.

lose an engine — in the heat of panic, probably means "one engine isn't working," not "an engine fell off the airplane." But not always.

lose the prop — to stop an engine inadvertently or to have an engine quit for any reason.

lost his keys — among "stewardesses," air-traffic controller "who fails to keep track of his planes," according to John Davis, *Buzzwords*. See also **park the car**.

lost the defrost — "de-icer malfunction," among "stewardesses," according to John Davis, *Buzzwords*.

low flying — see **balls**.

low observables — materials and structures designed to reduce aircraft radar and visual signature.

low rudder — the rudder turned toward the low side of an airplane while in a bank or roll. Also called bottom rudder. — Williams, *Casey Jones Cyclopedia of Aviation Terms*. See **bottom rudder stall**.

low tow — a hazardous situation in glider towing wherein the glider flies lower than the tow aircraft. See **high tow**.

lower 41 — electrical bay on Boeing aircraft, adjacent to the cargo compartment. So named because on the Boeing 707, the first aircraft to gather electrics into one place instead of scattering them around the aircraft, on the chart it was numbered "41."

Lufthansa — said facetiously to stand for "Let Us Fuck The Hostess As No Steward is Available." Another variation substitutes "stewardess" for "steward" and "aboard" for "available." Formerly spelled Luft Hansa.

mackerel sky — a sky rippled with high cirrocumulus. A longtime folk sign of weather on its way.

MAD — Magnetic Anomaly Detector, a sensing device used in anti-submarine warfare which detects underwater magnetic disturbance caused by the passage of a submarine, even underwater. Usually housed in the aircraft's long tail probe known as the **sting** or **stinger.**.

Mae West — 1. inflatable life vests worn by WWII aircrew, so named because, when inflated, they resembled some well-known attributes (body parts) of a popular cinema vamp of the time. See **yum-yum yellow**.

2. a parachute with the lines caught around it, so that the **silk bloomers** bloom out in two large bulges on either side of the twisted lines. See also **cigarette roll.**

mag, magneto — "a primitive ignition system discarded by the auto industry early in the 20th Century." — Baxter, *More Bax Seats*, 1988. A pre-flight must is to "check the mags."

making their numbers — making a report to the Intelligence Office or C.O., giving the number of E.A.s (enemy aircraft) shot down.

manometer — a mercury-filled U-tube that senses angle of attack, often used to deploy combat flaps or dive brakes.

Manson — among "stewardesses," "air-traffic controller," also known as **ATC** ("alcohol, tobacco, and cocaine addict"), according to John Davis, *Buzzwords*. Calling the person in the tower "Manson" suggests both the manipulative, controlling nature of the famous California convicted murderer and also expresses fear of the ATC's power over life and death.

Mapleflot — jocular name for a new national airline in circulation around 1991-92 when it was suggested that Air Canada and Canadian Airlines merge.

Marilyn Monroe design —see **area rule, Coke bottle design.**

markers — coloured target indicators dropped by **Pathfinder** (WWII bomber) crews that cascaded down onto the aiming point. Also coloured flares that floated on a parachute and were used when the ground could not be seen due to cloud cover.

mark one eyeball — "human sight. This term came into play during the moon landings, which were helped considerably by the mark one eyeball." — Dickson, *Slang!*

Mark Twain — "Since the Mitchells would bomb from minimum altitude the men removed the secret Norden bombsights. Captain C.R. Greening came up with a simple, twenty-cent [sighting alignment] device he called the 'Mark Twain' to replace the elaborate optical instrument; it was one of the great improvised weapons of the war." —Martin Caiden, "Doolittle's Raid on Tokyo," in *Fighting Eagles*, ed. Phil Hirsch.

marshaller — a **batsman**.

Marx brothers — among "stewardesses," according to John Davis, *Buzzwords*, a term for the control tower. See also **Manson, lost his keys.**

master bomber technique — "Pathfinders lobbed their markers in the middle, and for the first time the main force used the 'master bomber' technique that Gibson had started over the Moehne Dam; a **'master of ceremonies'** circled low directing the bombers by radiotelephone on to the choicest markers." — Brickhill, *The Dam Busters*.

May Day — international, universal distress call. From the French *m'aider* or *m'aidez*, two forms of the verb and pronoun meaning "help me," pronounced "muh-ai-day." See **pan**.

MBRAT — pronounced "embrat," code for "may be reached at."

meat ball, meatball — WWII Allied slang for the "Rising Sun" insignia on Japanese aircraft; a bright red ball. In a more descriptive phrase, "angry red meat ball." —Deighton, *Fighter*.

2. term applied to the large red circle sometimes painted on top of glider tow aircraft to help the glider pilot keep proper position during the tow. "Keep the tow plane's rudder tip lined up with the meat ball." Also known as **poached eggs**.

meat on the table — WWII USAAF expression referring to enemy twin-engine fighters, such as the Messerschmitt Me 110, known to be easier to shoot down than their single engine sisters. "His opponents were the deadliest our men ever encountered, for of the 28 German planes that [Robert] Johnson shot down in aerial combat, all were fighters. And of this number, only four were 'meat on the table." — Martin Caidin in the Foreword to Johnson, *Thunderbolt*.

mechanical cows — nickname for early Farmans. —Bowen, *Knights of the Air*.

medevac — medical air evacuation. See **casevac, mercy flight**.

MEL — Minimum Equipment List - a list of essential equipment, which varies with aircraft type and mission to be flown. If any of the items are missing or unserviceable, the aircraft is not considered **airworthy**. See **go/no go**.

mercy flight — an ambulance flight, usually a bush operation from a remote area, such as the Canadian north, to a place where a seriously injured or ill person can receive medical treatment. Typically, a matter of great urgency, made under difficult or hazardous flying conditions. See **medevac, casevac**.

Messer — a Messerschmitt fighter aircraft.

Messerschmitt Maytag — "light, low horsepower liaison type aircraft; check ship in **washout** flight." —*AAF: The Official Guide to the Army Air Forces*.

Mickey Mouse — bomb-dropping mechanism with rheostat, to activate, in order, release of heavy bombs, release of lighter, incendiary bombs, flash, and finally camera. WWII. —Doug Sample CD, Canadian Branch, Yorkshire Air Museum. Also, a twin-eared bomb release which allowed the dropping of one, or both, bombs or torpedos, such as in the Fairey Swordfish.

microburst — sudden downdraft in **windshear**, cause of a spectacular American Airlines crash at Dallas-Fort Worth in the late 1980s and some others since then. The microburst has been recognized as a specific phenomenon only in the last decade. It is now generally accepted that a great many earlier crashes, probably since manned flight began, were caused by a similar occurrence.

"First there is a downburst. Hitting the ground, it fans out like an inverted mushroom, and creates horizontal winds in all directions. On approaching a microburst during take-off the pilot flies into a headwind of up to 80 knots that rapidly increases his indicated airspeed; following his instinct or Flight Director, he pulls up to try to maintain climb out airspeed, but barely a mile away, the aircraft is slammed down by the vertical wind-shaft forcing it to descend. Losing indicated airspeed, the nose of the aircraft is now lowered in order to try maintain climb out speed...To complete the dramatic scenario, all within a minute or so, the wind changes to the opposite direction; the aircraft is in a nose-down attitude, and the sudden strong tailwind gives a further reduction in indicated airspeed, even to the point of stalling. As this happens, the **stick shaker** operates cueing the pilot to push the control column further forward in a desperate attempt to increase airspeed - which drives the aircraft into the ground. During approach, the problem is no less severe.

"The anatomy and nature of such downdroughts were discovered and explored by T.T. Fujita, Professor of Meteorology at Chicago University. He studied the JFK and similar incidents, and coined the name microburst for the elusive menace, a name that sounded precipitous, dramatic and sinister enough with a touch of science fiction flavour..." Barlay, *The Final Call*.

midair — a "midair." The unsayable gets said, but it is shortened. "A midair collision." Brits call it an **airmiss**.

Midnight — radio code word for "initiate advisory control (due to loss of radar)." —Canadian Forces Aircom Regulation Glossary.

Midnight Jones crews — "the unfortunates who arrived on the squadron one day and were told not to unpack because they were *on* that night — and were dead by midnight." WWII. —Dunmore and Carter, *Reap the Whirlwind*.

Mig Master — fighter pilot credited with **flaming** a **Mig**, during the Korean conflict. Also applied to certain US fighter aircraft. See Plane Names.

mike — microphone. Also, the phonic alphabet code word for "m".

Mile-High Club — fictitious organization of aviators claiming to have made love at least one mile above ground level. "After a hilarious trip down memory lane, (Bill) Lear and (Allen) Jones admitted to having started the infamous Mile-High Club in a DC-3 a mile above Lake Michigan. They had been flying in Lear's first autopilot airplane, and had left the controls to the new invention while they disported themselves horizontally with two young women in the passenger compartment." — Munro, *The Sky's No Limit*. The editors are of the opinion that Lear and Jones may not have been the first to do this, but probably were the first to talk about it.

The Toronto *Globe and Mail* of February 18, 1995, reports that a San Francisco enterprise called "Mile High Airways" offers a one-hour ride out of Hayward, California, in a single-engine Piper fitted out with a mattress, champagne, chocolates, strawberries, and an air-sickness bag, to couples who want to spend $275 to join the club. The story, by San Francisco *Chronicle* reporter Steve Rubenstein, discusses the difference between "turbulence" on this flight and on normal scheduled flights. ("Pilot plays Cupid on airborne lovers' joy ride.")

military pilots' oral rules — from WWII, specifically Bob Stevens' book of cartoons *"There I was...flat on my back*: 1. "A P-39 would tumble." 2. "The B-26 (Marauder) wouldn't fly on one engine." 3. "You couldn't safely bail out of a P-38." (Twin fuselages on either side of central cockpit, with lots of tail section on either side). "Note: actually, the T. O. said you were supposed to pull it up in a 'near stall,' jettison the canopy, walk across the stub wing, and *dive* over the boom!" 4. "If the P-47 hit 'compressibility' in a dive the stick 'froze'." "P. S. They also said this about the '38."

milk run — 1. like **piece of cake**, a bit of slang that has entered the general language so much that its origins in aviation are obscured, but Partridge agrees that it was first plane talk for an easy mission.

2. "a flight that has two or more stops." —Mary Manni, "Say What? A guide to airline slang," Air Canada's *Horizons* magazine, April 27, 1994

mill — 1. engine. 2. propeller, as in "get the mill spinning."

minimum sink — in glider talk, the speed in feet per second at which a glider descends in calm air at optimal airspeed.

missed approach — term used to describe a botched landing approach. Implies a **go-around**.

missing man formation — in this ritual of WWII, one aircraft peels off on a flypast to indicate that a comrade has been killed or lost (missing in action).

missions — typical WWII American airman's word for "took off, went to target, returned to base," or what the British and Canadian bomber crews called **ops** and the French and many English-speaking fighter pilots called **sorties**. The explosive spread of fundamentalist Christian mission efforts just after WWII suggests that a military culture spread beyond its boundaries; one of the major worldwide technological facilitators for this spread was the Missionary Aviation Fellowship, especially active in South America with small aircraft and experimental techniques for dropping supplies, Bibles, leaflets, etc.

Mistel — **Mistletoe.** German name for the unit formed when two aircraft are hooked together. Notably, near the end of WWII, in a desperation move, JU-88s would be packed with explosives and piloted by an aircraft attached above, then released over the target to form a diving bomb. See **Father and Son, Beethoven.**

MITO — USAF Strategic Air Command term meaning Minimum Interval Take Off, a procedure used to mass-launch all the aircraft on the base, also known as a **flush launch**. The procedure calls for a launch interval of 15 seconds between like aircraft and 30 seconds between unlike aircraft. See also **Alpha strike**.

mixture control — jocularly, "small red knob for making the cabin quieter." — Baxter, *More Bax Seats*, 1988. See **mess it up, engine stops.**

mock-up — a full scale model of the fuselage, without engine and wings, of a projected new aircraft, with emphasis on the interior. Engineers also make a "mock-up" of an aircraft; but what is tested in the **wind tunnel** is a "model."

modock — a pilot who always talks about what he is going to do, but never does it.

Moe — common "stewardess" term for the pilot, according to John Davis, *Buzzwords*. See also **Larry, Curly**.

moling — a daylight bomber raid, using cloud cover to conceal the bombers, from WWII RAF usage. —Dunmore and Carter, *Reap the Whirlwind*.

Molotov breadbasket — a metal container holding 12 or more incendiary bombs, which are scattered automatically after the container is dropped over the target. See **can**.

monsoon bucket — "this helicopter-borne water container is used for aerial bombardment of forest fires. The concept and early design are credited to Jim Grady and Henry Stevenson of Okanagan Helicopters Ltd., ...in B. C. Their early 1960s contraption used a 45-gallon drum with a basketball bunging a hole in the bottom! Modern versions are more sophisticated, of course, and claimed by other inventors. Elsewhere, called helibucket, or simply, water bucket." — Parkin, *WetCoast Words*. This method has been decisively outmoded by the scoop-and-drop Canadair CL 215 and CL 415T **water bomber**.

Montgolfier — French brothers, Joseph and Etienne, early flyers and balloonists. Term applied generically to all hot-air balloons in the early days.

moosh — the jail, the clink, in WWII military slang. —from Raff and Armstrong, *Plonk's Party*. See **cooler, digger, glasshouse**.

morning star — *morgenstern* (German), appellation given the protruding radar antenae of the WWII German Neptun radar, usually mounted on the nose of a Junkers Ju-88. See **stag's antlers**.

mort each other out — "(of Allied planes) to shoot each other down accidentally" — From John and Adele Algeo, "Among the New Words: Gulf War Glossary," *American Speech* (Winter 1991). "Mort" here sounds like a French word, but is actually also an obsolete English word, according to the OED.

Morton Fireball — sarcastic name for Morton Thiokol, after the Challenger spacecraft disaster, because they made the "o" rings that allegedly failed and caused the problem.

mosquito bomber — pesticide sprayer, such as the Stearman PT-13 Kaydet.

mosquito raid — a bombing or strafing raid made by a single aircraft.

mount — synonym for airplane. Also "steed," "charger," most often applied to fighters. Probably dates from the era when aircraft were operated by cavalry units — a natural extension of language.

moustaches — the nickname in French for the **canards**, small movable control surfaces such as at the front of the Mirage jet.

MRE — "Meal, Ready to Eat," officially; popularly known as "Meal, Rejected by Everyone." "The new dehydrated combat meals that replaced the heavier old C rations (that were referred to as C Rats)." — Halberstadt, *Airborne: Assault from the Sky.*

Mt. Boeing — Mt. Rainier, near Seattle, WA, "is so often used as background for Boeing pictures...that some people refer to the famed mountain as 'Mt. Boeing.'" — Cleveland, *Boeing Trivia.*

muck — thick, foggy, scuddy, stormy weather.

muddy — "said of weather that is turbulent." — Murray, "..Navy Fighter Pilots."

mud-mover — "Bomber pilot (as nicknamed by fighter pilots, UK source)" —From John and Adele Algeo, "Among the New Words: Gulf War Glossary," *American Speech* (Winter 1991). Also known as "truck driver" and "bus driver."

mufti — civilian clothes worn by military personnel. WWII.

mush — 1. (v.) to sink in a nose-high attitude in which the aircraft is barely responsive to normal control inputs.

2. British radar operators' term for distant, confused blips, too imperfect to be detected as aircraft or ground clutter, birds, etc. —Deighton, *Fighter.*

mush in — pull back and float in softly.

mushing — flying in a nose-high attitude, at low speed, barely above a stall.

mustard cluster — jocular WWII term for an imaginary award for poor bombing. —*AAF: The Official Guide to the Army Air Forces.*

NAAFI — (Brit.) Navy, Army and Air Force Institutes, operated Service Canteens.

NAAFI gong — jocular term for the Defence Medal, which was awarded to aircrews only for non-op time, and thus was said to have been awarded for the "service" of attending NAAFI bars at night.

—Doug Sample CD, Canadian Branch, Yorkshire Air Museum.

NACA — National Advisory Committee for Aeronautics, predecessor of NASA. (U. S.)

nape — Vietnam fighter pilots' term for a napalm bomb. —Col Bob Stoffey, *Cleared Hot!*

nap-of-the-earth (NOE) — term used to describe flying **on the deck**, close to the ground, following, and barely clearing, natural terrain. — Harding, *US Army Aircraft since 1947*. See **balls, contour flying**.

narrowbody — "aircraft with a single aisle and two cabins, First Class and Main Cabin." —American Airlines training materials. See also **widebody**.

narrow track — an aircraft with wheels placed closer together than others, so that it is in more danger of doing an occasional **ground loop**, e.g. the Messerschmitt Bf-109.

NASA — National Aviation and Space Administration (U. S.)

NASA wing — any of a number of new airfoil designs created by NASA research and development.

NATO codes — as morale boosters, or propaganda, or because they did not know what the other side called their aircraft, NATO uses code words to differentiate enemy plane types. See also **Allied code names**.

nav — navigation.

navaids — navigational aids. Any system or device that aids in aerial navigation, e. g. Loran, DME, rotating beacons, VOR, airways, markers, etc. Refers to both the ground systems and the instruments in the airplane.

navcoms — radios combining the functions of both navigational aids and communications.

Navy common channel — derogatory nickname among USAF pilots for the Guard emergency channel (123.5 or 243.0) because, according to them, Navy pilots use it most frequently.

needle, ball, airspeed — "Link trainer 'flight group", according to Bob Stevens, *"There I Was...."*. "Needle, ball, airspeed"' was the basic flight group — and keeping an aircraft straight and level on them in the **soup** was like walking a tightrope with a blindfold on."

needle-speedle-ball — WWII student pilot's term for Link trainer drill where emphasis was on the compass (or altimeter), airspeed and turn-and-bank indicator.

negatively raked platform — forward sweep wings, e. g. Junkers Ju-287.

nested — see **back-to-back, looped.** "The airlines consider nesting a 'misuse' of tickets." —John and Adele Algeo, "Among the New Words," *American Speech*, Spring 1994.

NET — "Not Earlier Than ('authorized abbreviation')." — Heflin, ed. *The United States Air Force Dictionary*.

new runway lights — name given to the strip lighting installed along the aisle floors in many airliners to help passengers find the emergency exits.

new undies — among "stewardesses," a "close call," a near-collision, a.k.a. **prayer breakfast,** according to John Davis, *Buzzwords*.

nickel raid — leaflet-dropping sortie (WWII) over France or Germany, often "crew-training graduate exercises."

night jump — obviously, a parachute jump made after dark, but among paratroopers, a ritual way to say that someone's jumps were always made with eyes tightly closed. —Pat Dudley. Some would say, "All my jumps were night jumps."

Night Witches — "When Nazi Germany invaded the Soviet Union, Communist officials accepted a thousand female volunteers from flying clubs to serve as fighter and bomber pilots. Despite their youth (aged 18 to 22) and Moscow's attempt to hide the fact it had resorted to enlisting women, the Germans had soon nicknamed the lethal fliers the Night Witches and put about the rumour they were castrated men. When it was apparent their success record equalled that of the best men's squadrons, writes Catrin Maelor of *The Sunday Telegraph*, the jokes about giggling schoolgirls disappeared. Night Witches flew under rotten weather conditions (some survivors are scarred from frostbite) and never wore parachutes." —"Social Studies," Toronto *Globe and Mail*, June 9, 1994.

niner — standard radio communications lingo for the number nine.

90-270 — like a **teardrop**, but with the manoeuvre shaped like a circle rather than an oval.

99s — female pilots, especially members of an aviatrix organization so-named because it had ninety-nine members at its first meeting. See also **Flying Seven.**

Nippon Clip-on — a removable wingtip used by All Nippon Airways on Boeing 747-400s for long range flights. "Since the airframe life is generally N takeoff-landing cycles and M hours of flying, ANA (and possibly others) flies without the wingtips on short flights, and when they used up a large number of cycles, they put the wingtips on and fly them on longer routes to get the most use out of the airplane." —"kls@ohare.Chicago.com" on an Internet electronic bulletin board.

no-fly zone — areas forbidden to hostile aircraft such as parts of Iraq and the former Yugoslavia. "You fly, you die."

No Goose, no Gander — key slogan in "El Al's classic advertisement for the carrier's nonstop transatlantic services in 1957, and referring to the Canadian airports at Goose Bay [Labrador] and Gander [Newfoundland] which were often used for refuelling." —**ICAO** 50th anniversary book.

no-handed — "Another relatively easy cost-reduction scheme would be to rethink aircraft design so that all parts are 'no-handed.' That is, there would be no left and right hinges or wing flaps or other control surfaces." — Rich, *Skunk Works.*

no joy — pilot's response to air traffic controller asking if he has spotted a nearby airplane to signal that he cannot yet see ("make visual contact with") the other craft. —Gordon Gilbert, editor, *Business and Commercial Aviation,* and a pilot. See also **Tally-Ho!**

non-rev — non-revenue passenger, as for example, flight crew being ferried to or from their assigned flight. **Deadheading, CONs.**

non-sked — non-scheduled, as in charter flights, **charter airlines.**

no smoking — dates back to the first recorded incident of a passenger caught smoking....aboard an Imperial Airways Handley Page HP-42 flight, Paris to London, in 1936. A Croydon court fined him ten pounds, a penalty considered severe in those days, presumably to deter others.

no-op — "a plane that will not operate; one that has been canceled." — Dickson, *Slang!*

norad, nordo — no radio.

Northworst — irreverent nickname for Northwest Airlines.

nose art — personalized drawings and other artwork painted on the nose of, mainly, military aircraft. Also called **war paint.**

nose gear, nose-landing gear — American term for what the British commonly call **nose-carriage, nose wheel.**

nose heavy — a condition caused by improper loading or improper trimming. Opposite of **tail heavy.**

nose-over — the turning of an aircraft onto its back, on the ground, by rolling over the nose, i.e. somersaulting.

nose wheel — jocularly, a device sometimes bent by the pilot, leading to "pilot's nose", usually bent right after the nose wheel.

no-show — ticket agent and flight attendant code for a passenger who has a reservation but who doesn't appear. A reason for overbooking, to try to fill all seats.

NOTAM — Notice to Airmen. Usually reports issued by officials or authorities to inform flyers that certain conditions exist, such as closed runways or restricted zones.

NOTAR — No Tail Rotor - a technology replacing the traditional helicopter tail rotor.

nugget — U.S. Navy term for a pilot on his first aircraft carrier cruise.

the **numbers** — term used when reporting time or pushback, liftoff, fuel aboard, **ETA**, etc.

Oboe — "a new way of radar pinpointing; two beams went out from England and crossed exactly over the target to let the pilot know he was there." — Brickhill, *The Dam Busters*. See **cat and mouse**.

Obogs — "On-board oxygen-generating system, the air supply in a jet plane (UK source)" —From John and Adele Algeo, "Among the New Words: Gulf War Glossary," *American Speech* (Winter 1991).

office — cockpit of an aircraft, also rear gunner's turret (WWII). —Doug Sample CD, Canadian Branch, Yorkshire Air Museum.

off-line — travel agent jargon for "travel on a carrier other than the one that sold the passenger the ticket." — Dickson, *Slang!*

Oh, Shit — a black box electronic counter-measure in the tail of the Blackbird SR-71 surveillance aircraft "called Oscar Sierra (the pilots called it 'Oh, Shit!' which is what most of them exclaimed when an ECM system activated at the start of an enemy missile attack)." — Rich, *Skunk Works*.

2. The most common phrase used by aviators in distress. According to aviation photographer Paul Bowen of Wichita, Kansas, "Those of us concerned with the spirit always say 'Holy Shit'".

oil burner — low altitude training route (military). —from Paul Turk.

oil canning — rattling sound caused by the bulging in and out of the fuselage sides in flight.

Old Ladies Home — the home airport. — Weseen, *A Dictionary of American Slang*.

olive branch routes — training routes flown by military jets.

onboard — equipment on the aircraft, such as an onboard computer.

on the deck — flying at treetop level. See **low flying, balls, hedgehop**.

on the clocks, on the gauges — instrument takeoff, flying, or landing. See **flying the cockpit**.

on the numbers — 1. a **three-point landing**. 2. touching down on the numbers painted at the end of the runway, i.e. near the **threshold**. 3. on schedule or, in engineering, "on specs" (constructed exactly to specifications). Also, on budget.

on the step — a way of describing the way the plane feels and handles after it has just finished its climb and leveled out at cruising altitude. That is, in cruise, when enough speed is attained so that the aircraft pitch attitude can be lowered slightly without causing the aircraft to descend.

2. in seaplanes, when the aircraft has climbed up on the "step" in its hull, on the take-off run, reducing drag from the water surface because of the reduced contact tension.

on top — above the weather, as when the aircraft has broken through and is flying above the cloud layer.

one-off — a ticket issued by an airline at a special bargain rate.

In engineering, this phrase has long meant a "one of a kind."

one-holer — single-person aircraft. See **two-holer, three-holer**. This phrase was originally applied to outhouses, and is borrowed by aviation from scatology.

one-oh-nine-itis — disorder suffered by some RAF pilots when engaging Messerschmitt Bf-109s, often leading to low oil pressure, overheated engines, and aborted missions.

one-third airplane, one-third helicopter, and one-third boat — not really an aircraft, but an Air Cushion Vehicle hovercraft used by USN in Viet Nam. —Mersky and Polmar, *Naval Air War in Viet Nam*.

one turning, one burning — engine on fire! This is a phrase normally used after the fact, but occasionally in the cockpit or on the radio.

one versus one hop — "simulated air battle between one student crew and one crew of instructors." —Murray, "Language of Naval Fighter Pilots." The phrase describes the degree of difficulty of the exercise. See also **two versus two hop, two versus unknown hop.**

one "G" roll — while flying the Boeing Dash-80, "swept-wing prototype for America's first jet transport, the 707, pilot Tex Johnston put it through a roll, to the surprise of the "shore-side assemblage" watching. "Boeing engineers on the plane said that Tex held it in a perfect one "G" roll. Tex explains that a plane doesn't know if it is flying level or on its back. It only responds to 'G' forces. At one 'G' in the roll, the stresses were no greater than in level flight." —Cleveland, *Boeing Trivia*. See **snap roll, flick roll.**

onion — exhaust bullet or cone on some jet engines, so called because of its shape.

open jaw journey — travel agent slang for a roundtrip in which the points of departure and arrival are not the same.

open skies — a phrase recently much in use to describe efforts to remove barriers to international air travel, such as restrictions on landing rights, varying **ATC** procedures and languages in use, and routings. Stuart Howard, in a rather political opinion, says it means "an open market in which the mega-carriers with the greatest market power will dominate the skies at the expense of weaker airlines." —**ICAO** 50th anniversary book. See **EATCHIP, fifth freedom, Single Sky for Europe.**

ops — typical British and Canadian WWII bombers' word for "operations," or what the Americans called **missions** and the French and many English-speaking fighter pilots called **sorties.**

Ops Wing — a small gold-coloured pair of wings with a large "O" in the centre issued by the RCAF (WWII) to indicate the wearer had completed a "tour of **ops.**" —Douglas Sample CD, Canadian Branch, Yorkshire Air Museum.

orange death — cynical nickname for aviation gasoline during WWI when being shot down often meant falling to a fiery death because of it. Also called **hell brew, infernal liquid, witches water.**

orbit — to fly in circles, or 360-degree turns, such as over a target or a downed aircraft. "We came in to the clear just before the target—I was not absolutely sure of it and I called for an **orbit** and was I ever unpopular with the crew and Jerry was sure throwing up a hell of a lot of heavy flak and we were predicted all the way around—picked up a few holes — on this run we let em go right in the slot." —F/O Sydney Wallis, "Logbook."

organized crash — jocular label for a landing in a difficult place, such as a rough surface, or too small an area, or, any landing.

ornithopter — a heavier-than-air aircraft supported in flight by planes to which a flapping motion is imparted.

osprey technology — U. S. aircraft manufacturers' research into methods to enable aircraft to take off vertically. —Los Angeles *Times* (February 23, 1995).

out airplaned — "(of a pilot) faced with superior enemy air power." ("Among the New Words," Gulf War Words supplement, *American Speech*, 1992). Jocularly (of an airplane)—met a superior aircraft.

overcontrol — to make large, erratic control movements that cause the aircraft to overreact. A "new flyer" syndrome. See **chasing the needle**.

oversexed, overpaid, and over here — derogatory rhyming description of the Yankee flyers in WWII, from the vantage point of the English pilots. —Jess Finney, AAC, San Diego. There was an American response: the British troops were said to be "undersexed, underpaid, and under Eisenhower." —Pat Dudley.

overrun — an extension of the runway capable of supporting an aircraft in the event of an aborted take off or an **overshoot**. See **stopway**.

overshoot — to run out of runway, either on takeoff or landing. Also, as a noun, applies to the area beyond the end of the runway. At some airpots this is a cleared area capable of supporting an overshooting aircraft. This area is sometimes known as the **stopway** and **overrun**.

over the shoulder toss — "low-altitude bombing technique," ("Among the New Words," Gulf War Words supplement, *American Speech*, 1992). See also **high toss**.

P — in the model numbers for early fighter aircraft, stood for "pursuit." After WWII, as Bob Parke has pointed out, they asked, "Who are we pursuing?" and changed the designation to "F" for "fighter". Hence a P-40 can be easily recognized as a WWII aircraft, by its letter designation. P-51 Mustang is an example of the change; it was known as the F-51 after 1947.

packs — air conditioning units on an airliner. Not to be confused with **PAX**.

paddle blade — squat wide propeller blade.

paddles — nickname for the LSO (landings signal officer) on a carrier, from the paddles he uses in signaling. —Coonts, *Flight of the Intruder*.

paint — radar reflection, signature. The significance of the reduced signature is at the heart of the success of the **stealth bomber**, **stealth fighter**.

Pakistan International Airlines — (PIA), "Perhaps I'll Arrive' and "Please Inform Allah" are irreverent nicknames.

pan — the paved area on which aircraft were parked. WWII. —Doug Sample CD, Canadian Branch, Yorkshire Air Museum.

2. International distress call meaning urgent but not as strong as **May Day**.

pancake in (v.), **pancake landing** (n.) — a wheels-up belly-flop landing. Also, should the landing gear collapse or fail to lock down, the aircraft is said to have "pancaked in." Doug Sample remembers that "pancake" could be used to refer to any landing, as well. Yet another sense of the word refers to a soft or gentle full-stall landing.

panic boost — full **through-the-gates** throttle, **firewall, buster**.

panther piss — "a special fuel additive...that ionized the furnace-like gas plumes streaming from the engine exhausts" of the Blackbird SR-71 surveillance aircraft; "the additive caused enemy infrared detectors to break up incoherently." — Rich, *Skunk Works*.

pantry — British term for an aircraft galley. The vehicle that delivers the meals is called the pantry truck.

pants — slang term for the wheel cowlings. —Berrey and Van den Bark, *The American Thesaurus of Slang*. See also **plus fours, tin drawers, spats, skirts, trousers**.

pants leg — the **wind sock**, the "cloth cone which indicates the wind direction." — Rose, *A Thesaurus of Slang*. See **tetrahedron**.

pantobase — a hull design which permits an aircraft to operate from water, sand, snow and ice, as well as land. The fuselage is sealed and fitted with heavily stressed land and water skis and wingtip floats. On takeoff from water, the skis are lowered and at 20 mph, they begin to skim the surface, acting as **aerofoils**, until flying speed is attained. Experimental and seldom used in practice. Tested on the Fairchild YC123E Provider.

paper airplane — design for a prospective new aircraft, as distinguished from "hardware," or airplanes actually built. —Bob Wohl, Canadair.

paperless airplane — an airplane designed completely by computer, such as the Boeing 777.

parachute approach — a landing approach during which the airplane is held just above stalling speed.

Paraffin Pete — nickname (generic) of the Duty Pilot at Yorkshire bomber bases during WWII, because "part of his job was setting out, or seeing that others set out, the flares which mark the runway at night-time." —from Raff and Armstrong, *Plonk's Party*. Of course the flares were powered by "paraffin" or "kerosene." See **flare path**.

parasite — an aircraft carried by or in another, the "mother ship." Experimental supersonic aircraft were frequently launched from another airplane.

parasite drag — incipient drag produced by any airframe because of air friction.

2. A whimsical term for a woman who drags her date into a jewelry store instead of an ice cream parlour. —Williams, *Casey Jones Cyclopedia of Aviation*.

parasol wing — wing mounted on struts or braces above, but not touching, the fuselage.

Parker Pen time — fake time in logbook entry. —McVicar, *The Grass Runway*. "When a pilot needs to show a certain number of hours in order to keep or renew his or her rating, sometimes it's Parker Pen time!"

park the car — "assigned altitude and heading by the air-traffic controller," among "stewardesses," according to John Davis, *Buzzwords*. "Hear about that **French kiss** at White Hair [O'Hare Airport] yesterday? The **Manson** musta forgot where he **parked the car**." See also **lost his keys**.

parrot — a Cold War device, issuing a transponder code identifying signal to enable the aircraft to be known as friend or foe. "Switch your parrot on," "Use Parrot Mode 3," etc. See also **squawk** and **choke the parrot**. In standard usage as a slang term for transponder. Also see **I.F.F.**

part-out — (v.) "That old crate is ready to part-out, for sure." "At some point in time an airplane is totally recyclable. The sum of the parts are worth more than the whole, which means you can sell the engines, the landing gear, the cockpit, for more money as components than you can the complete flying aircraft." — David Fowler, Pres., Evergreen Air Center, Marana, AZ., in *Airliners Magazine* May/June 1994.

pass gas — USAF slang term among KC-135 tanker crews for what they do: "we pass gas for a living." See also **boom**.

patch — airport, usually a grass strip.

pathfinder — aircraft sent ahead of main attack force to mark target or objective. WWII bomber slang. An old RAF saying has it that "after the pathfinders comes the shower," shower meaning "rabble" in British slang; an alternate meaning from Partridge's *Concise Dictionary of Slang and Unconventional English* gives "shower" as a military term, short for "shower of bastards" and earlier for "shower of shit."

patter — formulae used to help pilots, especially fledglings, remember sequences. One takeoff check from RCAF WWII goes: "H-T-M-P-C-G-S" or "Heat, Throttle, Mixture control, Pitch, Carb heat, Gas on, Switches on." See also **GUMP check, PUFF check**. Another, used in the Fairchild Cornell: "Tom Mix Fucks Four South Carolina Girls." (Throttle, Mixture, Flaps."......? —Williams, *The Plan*.

pattern bombing — tactic in which all bombers drop at the same moment their leader drops. Often used when formations had few experienced bombardiers available, the most senior and experienced being placed in the lead aircraft. —Deighton, *Fighter*.

PAX — insider shorthand for "passengers," from the ticket code. See also **self-loading cargo, talking ballast**, and **SOBs**.

pea soup — thick weather, especially fog. See **clag, soup**.

peel off — veering off away from a flight formation to make an individual dive at a target, or a landing.

pelican — French and Italian nickname for a water bomber, because of a main feature of the Canadair CL-215 fire-fighting amphibian which allows it to skim along the surface of a lake or waterway, scooping up water which it can then drop on a fire. See also "le **Canadair**." —from Catherine Chase. See also **Super Scooper**.

penalty box — "an inactive runway where an incoming aircraft waits until a gate becomes available." — Dickson, *Slang!*

penguins — 1. ground(ed) people who work in the aviation industry, e.g. ramp personnel, baggage handlers, from the point of view of those who fly. May be traced back to WWII air force (RAF, RCAF) slang, where pilots were known as **pigeons**. (Monty Berger). "It's a waste of time arguing with a penguin" — Dean Martin in the movie *Airport, 1970*.

2. "Aeroplanes used by both the French and American Air Services for initial training during the First World War. These machines could not fly, but ran along the ground while the pupil endeavoured to get some idea of how to control it by experiencing the 'feel' of the controls before being actually launched on his own in an aircraft that would fly." — Kent, *One of the Few*. Both meanings for this term are also given in Maurice H. Weseen, *A Dictionary of American Slang* and other dictionaries of the time.

people's plane — William T. Piper's description of the original Piper/Taylor Cub.

perspecs — see **clear perspex, plexiglas**.

Petit Bosche — name given a type of target used by British fliers in WWI for gunnery practice, basically, the outline of an enemy airplane drawn on the ground.

photobird — photo reconnaissance aircraft.

phugoid oscillations — the tendency of an airplane to alternately dive and climb, gaining speed in the dive until the added speed causes it to once again climb, slowing it so that it repeats the cycle once again. See **porpoising**.

PIA — See **Pakistan International Airlines**.

PIC — **Pilot in Command**, the one actually in charge of the flight, not necessarily flying the airplane.

pickle — a shot on a target, by a fighter firing a missile or a bomber making a drop. "One bad pickle from a pilot, one unexpected gust of wind, and you may be out of the running." "We could come out here and push the pickle button fourteen times. It would be nice to have fourteen **shacks**. But that's never been done before. Bombs go long and bombs go short." —Eric Hehs, "Gunsmoke 1993," *Code One* magazine, January 1994. Mr. Hehs explains that his research "gave the word a possible WWII usage related to the Norden bombsight.... F-16 pilots use it similarly, when they press the bomb release button, or 'pickling' a target, or, also, 'pressing the pickle.' The button is on the sidestick controller, which looks somewhat like a pickle, what with so many other buttons (cucumber bumps) on it."

picklebarrel — "shack or target in bombardier training." WWII. —*AAF: The Official Guide to the Army Air Forces*.

pickling — putting an aircraft in deep storage, usually entails spraying all windows to keep out ultraviolet rays, sealing all orifices, and topping off all gear boxes to keep air out.

piddle-pack — bladder relief bag, during the Gulf War.

piece of cake — easy operation. This term may seem to be in such general use now that it does not belong in this word list, but we have encountered it so persistently among WWII aviation people that we believe, with Partridge (who dates it from 1938), that it originated with them. Partridge also cites variants such as "piece of duff" (1943) and "piece of Gatow" ("gatow" echoing French "gateau," meaning "cake") from a 1948 cartoon (Gatow was a Berlin airfield vital to the supply operation in the Berlin Airlift). —Doug Sample CD, Canadian Branch, Yorkshire Air Museum.

pigs — anti-aircraft balloons, during WWII. — Dickson, *Slang!* See **sausage**.

pigeon — slang term for pilot, according to Monty Berger, during WWII, in Spitfire squadrons. See also **penguins**.

pilot — main term for the **driver** of an aircraft. See **ship**.

Pilot in Command — the one in charge of the flight, not necessarily flying the airplane, often referred to as **PIC**.

Pinball — "the act of pointing at one target/group with the intent of attacking another." —Canadian Forces Aircom Regulation Glossary.

PIO — "pilot induced oscillation," also known as **porpoising**.

pipe — another name for the **joystick**.

Piper — "as in Pied, who led all the children to flying." — Baxter, *More Bax Seats*.

Piper Cherokee — jocularly: flying Indian musician. — Baxter, *More Bax Seats*.

pipper — name for a gunsight dating from the days when the sight was a long tube mounted atop the panel and protruded through the windscreen, as in the Boeing P-26 **Peashooter**.

pip-squeak — early form of transponder **squawk** on RAF aircraft WWII. An **IFF** transmitter put out a constant signal recognizable as a friendly blip on **huff-duff** (radar) screens. —Deighton, *Fighter*. See also **strangle the parrot**.

pirep — Pilot report. Information passed along to the air traffic controller concerning flying conditions, such as bumpiness, ice, build-ups, etc., for the benefit of other pilots about to fly through the area. **Air rep.**

pit — the **belly** of the aircraft, where the luggage compartment is located.

pitch — 1. up and down movement of the aircraft.

2. the angle of attack of the propeller blades. See **fixed pitch.**

pitch-out — a sudden sharp turn away from the direction of flight, usually downward. See **bunt.**

pitot tube — inlet for air to drive such instruments as altimeter and air speed indicator. **Venturi tube.**

plane — a two-dimensional shape, with width and length but no depth, e.g. "horizontal plane," "vertical plane"; therefore in airplane jargon, a tail or a wing ("the tail plane"). Also "plane" means "airplane," but it would be overly precise to insist that an aircraft should be called a "plane," with apostrophe," and "plane" without the apostrophe be reserved for the draughtsmen and engineers.

planform — the arrangement of the wings, i.e. straight, swept, gull, etc.

plant it on — execute a very firm landing.

playing the piccolo — many modern aircraft have control buttons on the throttle lever and stick. Pilots can control radio, radar, and weapons by moving their fingrs. They call it "playing the piccolo." See also **pickle.**

playpen — WWII jocular term for the **cockpit**, a.k.a. **pulpit, dungeon.**

playtime — time spent in orbit over targets, "on station," during the Gulf War. —U. S. News & World Report, *Triumph without Victory*, 1992.

plexiglas — U. S. brand name for clear glass/plastic panels, See **clear perspex.**

plink — destroying, as in "target." "Tank-plinking was not our only mission." Letter to the editor, *Aviation Week and Science Technology*, (October 12, 1992), from Captains Gregory M. Anders and Les Morland.

plonk — WWII, a recruit. —Doug Sample CD, Canadian Branch, Yorkshire Air Museum.

plucking — airline terminal staff term for "referring to a Flight Attendant pulling the customer boarding passes at the terminal or aircraft door." —American Airlines training materials.

plug — fuselage section. **Barrel.**

plumber — RAF WWII, the squadron engineering officer. His motto: "Keep 'em flying."

plume — wake of hot air and gas emitted by a helicopter jetpipe or the wake from a jet engine at high power on the ground.

plus fours — wheel cowlings. —Berrey and Van den Bark, *The American Thesaurus of Slang*. Also **pants, tin drawers, skirts, spats, trousers.**

poached eggs. — slang term for Japanese WW II rondelles; also known as **meat balls.**

pod — engine nacelle, especially a jet engine.

pod-and-boom — descriptive of the traditional helicopter layout, a fuselage ("pod") trailing a long stabilizing "boom." — Harding, *US Army Aircraft since 1947*.

pogo — tanker boom.

pogos — "For taxiing and takeoffs, jettisonable twin-wheeled 'pogos' were fitted beneath the enormous fuel-loaded wings and kept them from sagging onto the runway while taking off. The pogos dropped away as the U-2 became airborne." — Rich, *Skunk Works*.

pogo stick — a tail prop, a device to keep a cargo aircraft with main deck cargo door aft of wing from tipping backward at the ramp (typical of B-727s, MD-80s).

2. what you call an aircraft which, during an air-to-air refueling procedure, breaks off and takes the boom with it, because of the protruding boom. Also known as a **unicorn.**

point of no return — the point along a line of flight from which it is no longer possible to return to the starting point due to fuel remaining.

poker — among "stewardesses," a passenger who pushes the call button a lot, according to John Davis, *Buzzwords*. See also **Etna, claw, faucet, sweater, tenpins.**

the **pond** — the ocean, Atlantic or Pacific. This term has spread to the general population, but seems likely to have started among pilots, with their lofty vantage point, just as they know the Rocky Mountains as the **rockpile.**

poop — slang expression for information or **gen.**

pop — slang term for Papa, the phonic letter P.

Popeye — "flying in clouds or a reduced area of visibility." —Canadian Forces Aircom Regulation Glossary.

popping up — see **pulling up.**

pop-up IFR — enroute change of flight plan, from VFR to IFR because of sudden adverse flying conditions. See **airfile**.

porpoising — pitching, as the image implies, losing control by increasingly violent oscillation of the nose of the aircraft, up and down, like a porpoise. In the crash of an F-22 advanced tactical fighter prototype, the "pilot retracted his landing gear and ignited his afterburners at roughly the same time and the plane's nose immediately began porpoising out of control." —*Montréal Gazette* (April 30, 1992). See **hunting, phugoid oscillations**.

Powder Puff Derby — an air meet at Reno, Nevada, featuring female pilots and smaller planes than at most air shows.

powerback — an alternate procedure to **pushback** for moving aircraft away from a gate. "If there are no obstructions behind an aircraft...and if the aircraft is a **narrowbody** aircraft, some airports authorize the aircraft to move backward on its own power." —"Airline Taalk," American Airlines Field Training/Ground Services materials, March 1993.

power brake and jet off-take — attributes of a 1973 Buick Electra that Ace, famous local black used-car salesman in Austin, Texas, was extolling to another soul brother. Another example of 'plane talk being applied to automobiles.

powered flaps— see **blown flaps**.

power egg — the engine and all supporting components (nacelle, cowling, etc.), considered as a unit and removable and transferable to another fuselage. Early version of what we now call "modular" technology.

powerplant — see **engine**.

power stall — usually a downwind landing, nose high, power on. The airplane wallows down the wind, just fast enough to remain airborne but slow enough to sink to earth. "Hanging on its prop." See **stall**.

PPF badge — gold coloured albatross worn on the tunic pocket of WWII bomber crews to indicate they served or were serving in the **Pathfinder** force. —from Doug Sample CD, Yorkshire Air Museum, Canadian Branch.

pranged — dented, "dinged," caused minor damage. British term, but also in use in Newfoundland, and then in Cape Breton, Nova Scotia. "Pranged her in the parking lot last night, did you?" Apparently over time the connotation of this word has weakened: "pranged" originally came from the Royal Air Force, where it meant "a crash," and, according to Murray

Kinloch (University of New Brunswick linguist), also "a successful strike on the enemy," as in a bombing run. A "wizard prang" was a total write-off (from Simon Petzold). To "prang in" was to crash (Bob Wohl, Canadair). An old RAF adage goes "when a prang appears inevitable, endeavour to strike the softest, cheapest object in the vicinity as gently as possible."

"Prang, meaning destruction, began in the days of fabric and wooden craft because with all those wires securing the wing struts 'prang' was the sound they made on crashing. In Bomber Command, the term came to mean the successful destruction of a target, hence 'wizard (highly successful) prang'". —-Hooper, *Pilot Officer Prune's Picture Parade*. See **bent plane.**

prayer breakfast — among "stewardesses," a "close call," a near-**French kiss** or near-**bingo**, according to John Davis, *Buzzwords*. See also **new undies.**

precision approach — an instrument landing. Usually the pilot must fly a precision course, from marker to marker, to line up properly with the runway. Old hangar flyers call "anything you can walk away from" a "precision landing."

preflight — more fully, preflight check or preflight hangar ("in colder climates, it is good to do preflight check in more nearly normal conditions" —Bill Stratton, International Liaison Pilot and Aircraft Association, San Antonio TX). See **walkaround**.

pressure cabin — the part of the aircraft, cockpit and passenger space, strengthened, sealed, and equipped to maintain normal ground-level pressure at high altitudes where the outside air pressure is lower, the air being thinner.

pressurize — to establish and maintain in part or all of an aircraft an air pressure higher than that found in the outside air at higher altitude. A joke among cabin maintenance crews is to get a rookie to perform a "air pressurization test," in which he walks backwards through the cabin gathering air into a plastic garbage bag, backs into the cockpit, ties the bag off, and hands it to the captain.

primer — see **wobble pump.**

probe-and-drogue — "Technique for mid-air refueling of a fighter jet by a KC-137 tanker plane, which drops a hose or drogue that connects to a probe from the jet (Canadian source)" — From John and Adele Algeo, "Among the New Words: Gulf War Glossary," *American Speech* (Winter 1991).

prop — (v.) to start an aircraft engine by manually **swinging the prop**.

prop blast party — among U. S. paratroopers, ritual celebration for someone's sixth jump, before which they were **cherry jumpers, snuffies**.

prop wash — US Navy joke expression, what they send a rookie off to find a bucket of. 2. **wake turbulence**.

propeller — jocularly, fan that keeps the pilot cool. Turn it off and watch him sweat.

Proud Bird, Golden Tail — official slogan for Continental Airlines, popularly parodied as "Proud Bird, Brass Ass."

pucker zone — bad flying conditions, in the clouds or in adverse weather.

puddle jump — a short flight. See also **hedge hop**.

puddlejumper — a term from the past, mostly, as applied to airlines, a small regional or feeder airline, like Trans Texas or Air Nova, often owned by a larger company. They used the smaller regional turbo-prop aircraft. Really the Mom and Pop airlines, which are, today, mostly replaced by regionals that are owned by the main airlines and fly larger aircraft.

2. a small, short range aircraft.

PUFF check — Australian mnemonic for Propeller, Undercarriage, Flaps, Fuel-pump: a check performed by pilots on short final to ensure all knobs and controls are at the right setting. See also **GUMP check, patter**.

puke — 1) "separate; follow one's own course (said of a fighter squadron during a **hop**. Usually used to describe a squadron's activity when it comes under attack by **bogeys**." —Murray, "Language of Naval Fighter Pilots." 2) Among military flyers, a term used to describe other pilots, such as "a bomber puke," or a "fighter puke." See also **jock**.

pukka — WWII term for "real, proper, or correct." "He gave us the pukka **gen**." From the Hindi, "pakka", meaning "substantial."

pull Gs — "withstand several times the force of the earth's gravitational pull when a fighter plane turns at high speed." — Murray, "..Navy Fighter Pilots."

pull out — manoeuvre designed to regain control of the aircraft after a spin or dive.

pull up — to increase power and bring the nose of the craft up to regain altitude, as after releasing a bomb. See also **popping up**.

pull your finger out — WWII military aviator's injunction to get on with it.

pulpit — cockpit of an aircraft, WWII. See also **office, playpen.**

punching holes in the sky — an expression for "flying," but with the connotation of its repetitiousness, the boring job of the airline pilot. "All we do is punch holes in the sky, one after the other." —from Bill Stratton, International Liaison Pilot and Aircraft Association, San Antonio TX ("Why I quit!") See **boring holes.**

punch out — "eject from a plane." —Murray, "Language of Naval Fighter Pilots." At Lockheed, during testing of **Have Blue**," code name for "the world's first pure stealth fighter," Bill Park, chief test pilot, got double hazard pay: "to Bill, even the opaque triangular cockpit is ominous, especially if he has to punch out." — Rich, *Skunk Works*. See also **drop out.**

pundit — name for aerodrome identity letters, flashing lights on ground in Morse Code (WWII, military aviation). —Doug Sample CD, Canadian Branch, Yorkshire Air Museum.

pundits and occults — RAF slang for **navaids.**

purple — "suspected of carrying nuclear weapons (air intercept code)" — From John and Adele Algeo, "Among the New Words: Gulf War Glossary," *American Speech* (Winter 1991).

Purple Heart corner — "outside plane in lowest flying element of bomber formation." WWII. —*AAF: The Official Guide to the Army Air Forces.*

pursuit — "The 1930s image of the brightly painted 'streamlined' biplane fighters, screaming as the wind rushed through their struts in 200 mph attacks, was gone; the colourful 'pursuit ships' had been replaced by drab green and brown monoplane 'fighters.'" —Vader, *Pacific Hawk.*

pushback — "action of aircraft being [mov]ed away from the gate by a tug or small truck, in order to move onto the taxi way." —American Airlines training materials. See also **powerback.**

pusher — rear engined aircraft, rear propelled aircraft. See **tractors.** During WWI the French air force decreed that all their aircraft be pusher-engined so that French anti-aircraft gunners would easily recognize them as friendly and hold their fire.

push-me-pull-you — generic name for aircraft propelled by both tractor and pusher engines.

put up a black — "risk adverse criticism from those in authority." WWII bomber slang —from Raff and Armstrong, *Plonk's Party*.

PWA — Pacific Western Airlines, nicknamed "Pray While Aloft."

Qantas — everybody knows this word, one of the few in English where a U does not follow the Q, but what does it stand for? Queensland & Northern Territories Air Service! (in Australia). See **Quaint Arse**.

Q-bay — "The heart of the U-2 were hatches in the equipment, or Q, bay that would house two high-resolution cameras......" — Rich, *Skunk Works*.

Queen Bee — senior administrative WAAF officer on a station. WWII.

Queen Mary — long low flat-bed trailer ("low-loaders") used to haul airframes, in WWII, especially crashed fuselages, to be rebuilt.

quick-turn burn — "Five-minute procedure for reloading an F-15 fighter with ammunition and fuel." —From John and Adele Algeo, "Among the New Words: Gulf War Glossary," *American Speech* (Winter 1991).

Quiet Birdmen — a secret society of pilots, extant since WWII, who wear an emblem and perform a ceremony when one of their number dies: all members present face west and say "He's gone west." David Hoover, who contributed the first item in this word-book, was member # 927, and has gone west. Paul Garrison calls it "a primarily social organization of pilots." —*Illustrated Encyclopedia of General Aviation*.

Q.C. — Quick Change, Quick Conversion. This refers to airliner cabin interiors furnished with seats mounted on pallets. They can be quickly on-or-off-loaded to change the seating configuration or to go to or from cargo configuration. Aircraft so equipped are usually designated QC, such as 707-320QC. **Reconfigure.**

Quaint Arse — nickname for Qantas, the Australian airline.

radar-friendly — "Easily detected by radar, like the B-52 bomber, as opposed to stealth planes (ironic)" — From John and Adele Algeo, "Among the New Words: Gulf War Glossary," *American Speech* (Winter 1991).

radial engine — piston cylinders encircling the central driveshaft. A round engine.

RAAF — Royal Australian Air Force.

radar — British first called it RDF, Radio Direction Finding, then Radio-Location, eventually "radar," for "*ra*dio *d*etecting *a*nd *r*anging." WWII. See **huff-duff.** —Deighton, *Fighter.*

radome — dome-shaped housing for a radar antenna.

RAF — Royal Air Force, the British air military division. To it were attached during World War II, the remnants of the Polish Air Force and the Free French air force, which survived to carry on the battle against the Luftwaffe. Known also as the "air works." The increasing importance of the air war (some say it became during WWII the most important part of the military) is reflected in the fact that every branch of the American military force had an air component—the Army, the Marines, the Coast Guard, the Navy. See **WAAF, WASP, AAC, AAF.**

2. Reform Air Force, a name applied to Preston Manning's aircraft, donated by supporters of his Reform Party in Canada.

RAF radio lingo in WWII — some standard items, in addition to **bogey, bandit, tally-ho,** were also "liner" for "cruising speed," "buster," for "full throttle," and "pancake" for "come home and land." —Deighton, *Fighter.*

rag wing — fabric-covered aircraft.

rain on the plane — aircraft engineers' word for "water vapor that has condensed out of the cabin air on contact with the cold skin of the plane," and will fall on passengers unless prevented by, as in the Boeing 777, "a small green doughnut that absorbs the moisture at intervals along the plane and prevents such internal showers." —Sabbagh.

raise the dead — said of aileron trim, helpful in maintaining straight and level flight with one engine out, i.e. raise the wing with the dead engine.

ramp rash — same as **hangar rash** but it happens outdoors.

ramp rat — anyone who works outside, at the airport, refueling, directing traffic, handling baggage, etc. In past times, sometimes applied to the kids who hung around airports hoping for a **hop.**

ramp strike — naval air expression for a too-short carrier landing.

ramrod — WWII fighter pilot's expression for a bomber escort mission. According to Monty Berger, a "short range bomber raid." See **little friends** and **big friends.**

ranger — In WWII a daylight raid on the Continent.

rat — "ram air turbine," a device that swings down automatically on engine power loss, and through converting the energy from the air forced through a small turbine to hydraulic energy, gives the pilots control over rudder and other "control surfaces"; it was because of a **rat** that the **Gimli Glider** was able to land safely after running out of fuel. —John Wheeler, Boeing public relations.

RATO — rocket-assisted takeoff.

RATOG — British term for "rocket assisted takeoff gear."

RCAF — Royal Canadian Air Force, before consolidation in the late 1960s into the Canadian Forces.

read — phrase used to mean "hear", as in "I read you."

read-back — phrase used to mean "repeat the information as received," as in "I need a read-back of my last clearance."

ready deck — an aircraft carrier deck prepared to retrieve landing aircraft. Also known as a **green deck**. —Coonts, *Flight of the Intruder*. See also **clear deck, foul deck**.

recce — pronounced "recky", means reconaissance; usually said of an aircraft, sometimes of a mission or assignment. **Recon**.

reconfigure — change the interior seating arrangement. Typically, adding or deleting first class, business class or coach class or going to or from cargo configuration. Some aircraft have seats mounted on pallets which can be quickly onloaded or offloaded as required. They are usually designated **QC**, as in 707-320QC.

recover — USN expression meaning "the landing of aircraft on a carrier." "By the time the Ranger had recovered her aircraft...." —Mersky and Polmar, *The Naval Air War in Viet Nam*.

Red Arrows — Royal Air Force aerobatic team. Other national aerobatics groups include Canada's **Snowbirds**, the U.S. **Blue Angels** and **Thunderbirds**.

redeye — a late-night flight, e.g. from coast to coast, as in "I came in on the redeye from L.A.," or, for that matter, any very early morning flight.

Red Flag — a dogfighting school, like **Top Gun**.

red hats — in parachuting, the **riggers**.

red-on-blue — "you're going the wrong direction." The opposite of **blue on blue**, with "red" signalling stop, turn around, go back. Also used to describe having put the compass on reverse, which, if you forget, sets you going the wrong direction. —McIntosh.

redout — "It's much like blacking out, except that everything goes red, and your vision goes." Often occurs when "you're going downwards, inverted, at a very high rate of knots." "You kinda see red through your eyes rather than having that black out feeling of losing consciousness." — Williams, *The Plan*. **Blackout, whiteout, greyout, brownout.**

Red Shirt — U.S. Navy term for an aircraft carrier crewmember in charge of aerial ordinance, so named because of the color of his shirt. See **Blue Shirt, Brown Shirt, Yellow Shirt.**

red streamer, red flag — signals a student pilot, red pennants being displayed prominently on the aircraft.

Redtails — U. S. Army Air Force black squadron, the **Tuskegee Experiment,** trained during WWII to fly convoy duty.

recip — short for reciprocating engine.

Regia Aeronautica — name of the Italian Air Force under Benito Mussolini during WWII.

regional jet — new concept, design, and aircraft of the 1990s. BAC calls a version of its BAe-146 the four-engine RJ-70, and there is also the two-engine Canadair Regional Jet. The concept enables airlines to bypass the huge hub airports (O'Hare, Atlanta, LaGuardia, Memphis, etc.) and fly smaller numbers of customers directly between cities on **long thin routes** with fuel-efficient, two-pilot aircraft.

RESCAP — Rescue Combat Air Patrol. See **SARCAP, BARCAP.**

respect armament — defensive armament meant to deter attackers by keeping them at bay.

rev, revving up — accelerate the engine before taking off. See **run-up.**

rhubarb — low level attack by pairs of aircraft on road traffic, etc. WWII. —Doug Sample CD, Canadian Branch, Yorkshire Air Museum. "Whence came the name, or why, or what it meant to Hurricane jockies remains something of a mystery, for the word, applied to action, is an American one and is baseball writer slang for an argument or jawing match between players and umpire, or between players of one team and players of the other, with the umpire usually in the center. This term was probably derived from the theatrical direction to mobs to keep saying, 'rhubarb, rhubarb, rhubarb' over and over again, the resulting sound being that of angry or protesting crowds of people.

"But how this came to be applied to intrusions into France and the Low Countries by single Hurricanes or flights of three to six is one of the mysteries of wartime nomenclature and pilot vocabulary." — Gallico, *The Hurricane Story*.

Larry Forrester, in *Fly For Your Life*, describes it thus: "'Rhubarb'" was the name given to a type of operation specially designed for cloudy conditions. Singly or in small groups, the Royal Air Force's fighters darted across the Channel, dived out of the overcast to shoot up selected targets, then climbed back into the murk before the Messerschmitts appeared. These activities, which demanded considerable skill and experience in navigation and marksmanship, had been standard 'rainy day' procedure for more than eighteen months - part of Air Chief Marshal Sholto Douglas's policy of 'leaning forward into France.'"

Finally, *Time* magazine of March 22, 1943 identifies it also as "British flyers' slang" for "a target of opportunity" — when "a fighter pilot flies low over France, strafing whatever he finds—trains, troops, airdromes—he is on a 'rhubarb'."

rhubarb roll — "It was during this mating time of man and machine that (Larry) Robillard gained a nickname. A Polish pilot, unable to pronounce Larry's surname, found it easier to say 'Rhubarb.' Later on, Larry would unintentionally invent a new aerial manoeuver that would bear his name. It was known throughout the war as the Rhubarb Roll." —Munro, *The Sky's No Limit*.

ride — 1) his or her final test flight after dual instruction. Also, any **check ride.** 2) term to describe the smoothness or roughness of air due to **chop** or turbulence, such as "How's the ride at 35000 feet?"

ride report — a **pirep** on turbulence conditions at various altitudes.

riding side-saddle — scornful way to describe the flight engineer, who sits at right angles to the nose in the cockpit. —McVicar, *The Grass Runway*.

riding the ranges — early transcontinental flying done by "following clearly defined radio beacons audible to pilots." — Beaudoin, *Walking on Air*.

rigger — 1. British term for airframe mechanic. See **fitter**.

2. in parachuting, those assigned to inspect and pack the parachutes and do the "rigger check" of equipment for each man on the aircraft

just prior to the jump. —Pat Dudley. See **red hats.**

roach coach — "flight to the tropics," in commercial aviation. — Dickson, *Slang!*

roach rats — among "stewardesses," "coach passengers," according to John Davis, *Buzzwords.*

rock and roll — airworthiness directive for older Cessnas prone to water-in-fuel contamination. Pilots were directed to shake the wings vigorously during **preflight** to dislodge any water caught in fuel bladder low spots.

the rockpile — the Rocky Mountains, to pilots, such as for Aspen Air, who fly over them a lot. —John Race, Canadair Challenger pilot.

rodeo — WWII Allied expression for a fighter sweep over enemy territory with the intention of luring enemy aircraft up for a fight.

Roger — jocularly, the most popular name on the radio.

rogue — a pilot who ignores regulations, especially one who flies IFR without maintaining contact with **ATC** while in controlled airspace, or one who is apt to **bust clearance.** Sometimes called a "luge-cannon" or "renegade."

roll — bank the aircraft through a complete rotation while flying straight ahead. Various manoeuvers have specific names. The following are from Williams, *Casey Jones Cyclopedia of Aviation Terms*: aileron roll - the revolving of an airplane around its longitudinal axis solely by the use of the ailerons, also called a "slow roll;" barrel roll - a complete 360-degree roll about the longitudinal axis; snap roll - a fast, complete roll about the longitudinal axis without change of direction or flight level; double snap roll - two snap rolls, one following the other without pause, calling for a fifth more flying speed than the single roll; dutch roll - the irregular movement of an airplane in flight due to gusts that cause the wing tips to rise and fall alternately in a circular motion.

rollers — British term for **touch and go, circuits and bumps.**

rollout — first public showing of a new airplane.

Rolls-Royce — for a list of the World War I and II planes (e. g. Spitfire, Lancaster, Hurricane, etc.) powered by engines made by this automobile manufacturing firm, see Mike Fox and Steve Smith, *Rolls-Royce: The Complete Works: The Best 599 Stories About the World's Best Car*, together with many stories of the interface between these cars and the world of planes. These include the stories of early VTO (vertical takeoff) craft (the **flying bedstead**), the first direct flight across the Atlantic in 1919 by Alcock and Brown, Royce's ability to deduce the speed of the aircraft from the pitch of the engine noise (from the ground!), the flight-testing of the Messerschmitt with a Rolls-Royce engine, the first turboprop and turbojet airliners, and much more. See also **crash**.

Roman candle — see **cigarette roll**.

roof — U.S. Navy term for an aircraft carrier's flight deck, since most of the ship was located below the "roof."

roof-hook — the most colourful of the synonyms for a dramatic launching device to fly Piper Cubs off and land them on an LST during WWII. More commonly known as "Brodie gear," for the inventor, James H. Brodie, a lieutenant in the U.S. Army. The device was "a wire cable, tightly strung between steel posts and outrigger booms along the port (left) side of the LST. A nylon sling attached to the cable on rollers dangled down a few feet, and the Cub had an arrestor hook fixed to the cabin roof by which it hung down from the sling." In a wind, the launch crew and pilot would have to consider what they called the "angle of the dangle" of the nylon sling. —Gordon Baxter,

"Bax Seat: Cubs at Sea on an LST," *Flying* (September 1994). The first LST to be rigged for this launch became known as the "Brodie ship." The "Brodie" name is not to be confused with that of Steve Brodie, the first man to jump off the Brooklyn Bridge in 1886.

ropey — scary, frightening. WWII military use.

rotary engine — piston cylinders revolve around a central drive shaft. Mainly WWI.

rotodome — large saucer-shaped radome mounted above the fuselage of some early-warning surveillance aircraft, rotates to "look" in all directions.

rotor — 1) a horizontal air eddy, usually to the leeward of a mountain or range, often strong enough to roll any aircraft flying through it. Suspected cause of the crash of United Airlines Flight 585, which **augered in** approaching Colorado Springs in 1991.

2) In helicopters, the tail stabilizer and/or the rotary wings.

rotor-head — "Helicopter pilot" — From John and Adele Algeo, "Among the New Words: Gulf War Glossary," *American Speech* (Winter 1991).

round robin — a cross-country flight in which the aircraft returns to the starting point without having landed elsewhere.

roundel — a round national insignia normally painted on the fuselage sides and on upper and/or lower wing surfaces. Usually three rings of colour. Best known as the blue and white BMW symbol, supposed to represent a spinning airplane propeller. BMW built the engine for the German Foche-Wulf fighter during World War II (Mercedes powered the Messerschmitt).

round engine — radial engine with piston cylinders mounted in a circle around a central crankshaft (Pratt & Whitney Wasp, Wright Cyclone, P & W 4360, etc.) —from Paul Turk.

R.P.M. — revolutions per minute, referring to the engine or prop, measured by the tachometer, or in Britain, the "rev counter." Jocularly, initials of a large corporation that builds tachometers.

put the **rubber on the ramp** — transform a "preliminary design concept,...making it practical so that it could actually fly....Baldwin had to make certain that the shape's structure was sound, ...determine its radius, its thicknesses, its ability to withstand certain loads, the number of parts. 'Baldy' would put the rubber on the ramp." — Rich, *Skunk Works*.

ruddervator — moving flight control surface of **butterfly** tail which combines functions of rudder and elevator, as in the Beech Bonanza. See **V-tail**.

Ruhrflacken — imaginary, jocular name for bomber destination, deep in Germany, with lots of antiaircraft guns. — Stevens, *"There I was."* The name may have been made up, but the destination and the guns were real: Doug Sample, a **tail end Charlie** on Handley-Page Halifax bombers out of Yorkshire, said of the Ruhr industrial heartland [Dortmund, Essen, Bochum, Düsseldorf] "that was my target logbook." **Happy Valley.**

run-in — said of new or overhauled engines. The purpose of the procedure, performed on a test stand, is to check engine power, ensure no fuel, oil or induction leaks, properly set the fuel injection system and initiate the **break-in** process. "During run-in we're trying to point the engine in the right direction. We're not breaking-in." — Tom Rempe, test cell supervisor at Mattituck (NY) Aviation in *Air Progress Magazine*, Oct. 1994.

2. British expression for the landing approach.

run-up — reving the engines to full throttle, testing the **mags**, before takeoff to ensure all systems are go.

runway — usually paved, but always "prepared" surface in an airfield where aircraft takeoff and land. Jocularly, the place where exotic stewardesses start the **airstrip**.

runway denial device, JP233 — (Brit.) "bomb that scatters clusters of cratering bombs over a wide area to destroy air base runways," ("Among the New Words," Gulf War Words supplement, *American Speech*, 1992).

RX RX — "Rush," in ticket agent and flight attendant slang. —from John Cavill.

SAAB — Svenska Aeroplan AB, a Swedish company that builds automobiles and also airplanes, including jet fighters. .

SAAF — South African Air Force (WWII).

Sabena — the Belgian airline. Jocularly said to stand for "Such a Bloody Experience Never Again." See **SAS**.

S.A.B.S — stabilizing automatic bomb sight, improved RAF WWII bombsight for precision bombing.

Sabre jock — pilot of a Lockheed F-86 Sabre jet fighter. Similarly applied to the driver of any type of aircraft such as, "Hurricane **jock**," "Cessna **jock**," etc. See also **puke**.

SAC — Strategic Air Command, global USAF task force to counter the Soviets, during the Cold War. See **TAC**.

safe life — time, determined by testing, after which a component should be replaced, even if not apparently defective. Often shortened to "life." Said also of the entire aircraft.

safety — "Very few pilots make it to retirement. If a mid-air, a dangerous approach, or a faulty instrument doesn't get you, perhaps a six-month proficiency check or physical will." —Capt. X, *Safety Last*. "If the bird don't get you, the **regs** will." "Keep your head on a swivel."

SAHSA — Stay at Home, Stay Alive is jocularly said to be the meaning of these letters, which refer officially to Servicio Aereo de Honduras Seguridad Asegurado, the Honduran airline. .

SAM — surface-to-air missile.

2. Also, Sociedad Aeronautica de Medellin, Colombian airline, said popularly to stand for **Sientese, Agarase y Matase**, which see.

SAR — Search and Rescue. See **SARCAP, RESCAP, BARCAP**.

SARCAP — Search and Rescue Combat Air Patrol.

sardine tin — a torpedo bomber. —W. H. *What's the Gen?* Torpedoes were known as **tinfish**.

SAS — "Sex and Satisfaction", "Sex after Service" and "Same as Sabena" are irreverent nicknames for Scandinavian Airlines System. See **Sabena**.

SATS — Short Airfield for Tactical Support. "The diminutive A-4Es used by the Marines at Chu Lai operated from an innovative set-up which was essentially, a land-based aircraft carrier.....the airstrip employed a catapult and arresting gear, just like a carrier." —Mersky and Polmar, *The Naval Air War in Viet Nam*.

sausage — WWI British slang for an enemy observation balloon, which was often of elongated shape, and was considered a tasty target. See **pigs**.

saw-tooth — step in wing leading edge where chord increases. Same as **dog-tooth**.

say again — term used in radio communication to mean 'repeat.' Also applied as in "say altitude," "say heading," "say position," etc.

scan — pattern of scanning cockpit instrument readings, especially at night or in instrument flight. Also used to describe checklist procedure, such as **GUMP check, PUFF check**.

Scare Ontario — irreverent nickname for Air Ontario.

scarf up — "from Vietnam. 'To grab, rescue, capture,' as 'scarfed up by the Jolly Green Giants.' Unfortunately, can also be used as 'scarfed up by the VC.'" — Stevens, *"There I was....."*

Schneider Trophy — offered for fastest flight during 1930s.

schräge musik — vertical firing 20mm cannon of German fighters, mounted so they could fire upwards. Appeared late in 1943. German for "oblique, slanting, sloping music." —Doug Sample CD, Canadian Branch, Yorkshire Air Museum. Sample recalls that he heard the term only after the war, and always from Allied personnel, but knew at first hand these upward-firing antiaircraft guns. Also translated as "jazz music." See **Ruhrflacken**.

schwarm — formation often used by German fighter pilots during WWII, consisting of several "pairs," a leader, and a wingman, with pairs in loose formation, at slightly different altitudes to maximize collision-avoidance and mutual protection. Known to Allies as the **finger four**.

scope — the face of the cathode ray tube of the radar, or in a **glass cockpit**, the face of the digital readout tubes, as in "I have that traffic on the scope", or, "I'm showing an engine fire on the scope." See **fish finder, Stormscope.**

Scotch mist — to be suffering from bad eyesight. WWII military aviator slang. —Doug Sample CD, Canadian Branch, Yorkshire Air Museum.

scoresheet — the record of an individual pilot's victories or bombing missions, painted on the side of his aircraft.

scramble — code/slang word for "time to go," when fighter pilots and crew are mobilized to meet a bomber attack force.

scrambled eggs — gold braid on the cap visor of a Group Captain or higher ranking officer (WWII). See **fruit salad.**

screened — among WWII bomber crews, to have completed a tour of operations (25-30 missions, bombing **ops**), followed by leave and six months teaching. Third tour was voluntary. —Douglas Sample CD, Canadian Branch, Yorkshire Air Museum.

screw sked — "crew scheduling", according to commercial airline crew slang, because "they always screw us up." —anon. Air Canada employee.

scrub — to interfere with airflow in such a way as to destabilize a smooth landing approach or in a turn. After deploying the spoilerons, ... "caused a rapid but smooth deceleration. Near the threshold, I executed a 180 degree turn. The aircraft responded nicely and the nose gear did not scrub." —David Hughes, Pilot Report in *Aviation Week and Space Technology*, Farnborough issue, September 7, 1992. Also to be **grounded** or **washed out.**

scrub board — the medical examining board which passes judgement on pilots' physical qualifications. If you fail, you are **scrubbed.**

scrubbed — WWII term for operation cancelled, also now any cancelled flight or failed attempt. "To cancel or to back out of a countdown." — Dickson, *Slang!*

scud — small, low, wind-driven clouds, mist or fog. "Only the advent of more technically capable aircraft, able to fly right through the scud and bomb accurately with little or no visual referents, would overcome the tremendous problems posed by weather." —Mersky and Polmar, *The Naval Air War in Viet Nam.*

Scud buster — "The Patriot air-defense system; a soldier working in that system; an F-15E **Strike Eagle** jet." — From John and Adele Algeo, "Among the New Words: Gulf War Glossary," *American Speech* (Winter 1991).

scud run — a flight necessarily made at low altitude because of clouds. See **hedge hop**.

SEAL — "Naval Sea Air Land commando force; a member of it," ("Among the New Words," Gulf War Words supplement, *American Speech*, 1992).

seamless travel — "what airlines hope to give passengers through codesharing arrangements with other airlines." —**ICAO** 50th anniversary book.

seaplane — though often used synonymously for **flying boat**, a seaplane is more usually one with floats or pontoons, a **float plane**, while a flying boat carries its flotation capabilities in its hull-configured fuselage. "An aircraft that carries its own airport on its bottom." —*Flying*.

seat-of-the-pants flying — old-timers' term for trusting your instincts and the "feel" of the aircraft when instrumentation was nonexistent, unavailable, or unreliable.

Second Dickie — a new bomber pilot (WWII) who flies with a seasoned crew before taking his own crew on bombing operations. —from Doug Sample CD, Yorkshire Air Museum, Canadian Branch.

see and avoid — adage for "see and be seen."

seeing-eye set — RAF WWII slang for onboard radar, probably first applied to night fighters.

seed — "plane that flies on the interior of a fighter squadron formation surrounded on four sides by other planes." —Murray, "Language of Naval Fighter Pilots."

self-loading cargo — insiders' airline industry slang for "passengers." **PAX, SOBs, talking ballast.**

senior mama — number one stewardess or flight attendant. Also known as the "in-charge." —Mitchell. See also **cockpit queen**.

sequence — term used by **ATC** to indicate position in the landing or takeoff pattern. As: "Expect **vectors for sequence** runway 2-4 Left."

SERE — "survival, evasion, resistance, and escape." Key words for a downed military pilot.

serviceable — ready-to-fly aircraft: "a pilot can sign for it and fly it away." —Deighton, *Fighter*. See **available**.

sesquiplane — biplane with a lower wing very much smaller than the upper plane. Sometimes called a "one and a half" or a "half-plane."

severe clear — perfect weather, and rhyming slang. —Gordon Gilbert, editor, *Business and Commercial Aviation*.

shacks — direct hits on a target. "We come out here and push the **pickle** button fourteen times. It would be nice to have fourteen **shacks**. But that's never been done before. Bombs go long and bombs go short." —Eric Hehs, "Gunsmoke 1993," *Code One* magazine, January 1994.

shakey-do — a mission plagued by serious problems. A **hairy-op**. —Doug Sample CD, Canadian Branch, Yorkshire Air Museum.

Shaky Jake — nickname for the Jacobs radial engine.

shark bait — among "stewardesses," a term for the life jacket, according to John Davis, *Buzzwords*. The term echoes but does not necessarily derive from the custom among WWII flyers of referring to the yellow colour of the **Mae West** life jackets as **yum yum yellow**.

shear — **windshear, microburst**.

shelter first, then signal — training adage, advice on priorities after an air crash. —Robert C. Yeager, "Mercy Flight Missing," *Readers Digest* (February 1995).

keep the **shiny side up and the greasy side down** — trucker's slogan, but also a pilots' oral rule. American Airlines pilots say "keep the blue side up."

ship — fairly common term for an aircraft. It is short for **"airship"**, which later became applied mainly to dirigibles. By so naming the airplane, humans seem to have been trying to associate powered flight with the solidity and stability of ocean-going vessels, which, despite the Titanic and Empress of Ireland sinkings, seemed trustworthy. This cluster of connotations was continued by naming the main "driver" of an airplane the "pilot," who was, after all, a specialist in bringing large vessels into port, through shallows and sand- or rock-bar-strewn channels, brought on board just outside harbour and dropped off after the ship had left dangerous waters. The same seems to be behind the term "captain" for the **pilot-in-command** of an aircraft.

Bruce Callander supports this reading of the origins of airplane slang by noting that, like mariners, early aviators' "three dimensional world well understood the concepts of roll, pitch, and yaw. Like boats, flying machines had rudders, ribs, spars, stays, and screw propellers. For a time, the new vehicles were called aeronefs (air ships) and their operators were known as aeronauts (air sailors). One reporter, trying to describe how an early pilot banked his machine in flight, wrote, 'The navigator threw over his helm.'" —"Jargon of the Skies," *Air Force Magazine*, October, 1992. Another corroboration comes from Rick Drury's "Flightlines" column in *Airways*, on the uniforms worn by pilots: "Since the airlines asked passengers to actually leave the earth and enter the unknown world of the air far above, it was imperative that figures of solid authority be in command and garbed in uniforms that exuded power and confidence. So the uniforms began to become almost nautical in style with huge stripes on the sleeves to denote the captain, the co-pilot—eventually redesignated first officer—and the flight engineer, later called the second officer. Even the names were nautical in tone." (Vol. 1 No. 6, January/February 1995).

Barry Lopez also points out that "the heritage of oceangoing vessels is preserved in the language and some of the design of modern airplanes. Pilots frequently call the plane a ship; its fuselage, a hull. Its interior space is divided into decks that extend fore and aft. The captain might refer to starting an engine as turning a wheel. He steers the plane on the ground with a tiller and speaks of docking the ship, after which, on a freighter, cargo is always taken off the main deck on the port side....A rudder in the plane's vertical stabilizer changes its course. Waterline numbers stenciled on the interior of the hull indicate height above the ground. Sailboat fairings taper engine mounts into wings that bear green running lights to starboard, red lights to port."

shit-hot — a noun, "the most elite kind of fighter pilot." — Murray, "..Navy Fighter Pilots." Also, and probably first, an adjective. Also said, in the code form, **Sierra Hotel**. Murray comments, "'Shit-hot' reflects linguistic creativity in the morphological metathesis of 'hot shit.' Interestingly, only the form of the phrase has changed; its meaning in both instances is (ostentatiously) skillful'."

shock watch — "a device affixed to a shipment that will register **g forces** or impact sustained by the shipment while in transit." — American Airlines training materials. See also **Tip-n-Tell Label**.

shoot — catapult launch of an aircraft from a carrier deck. See **cat shot.**

shoot (off a watch) — "energetically relate the complete story of a **hop** during free time, while being completely oblivious to all else.....'shoot' ... is semantically identical to that in 'shoot the breeze,' ....from which it may be derived." —Murray, "Language of Naval Fighter Pilots." See also **line-shoot, start the lamp swinging.**

short snorter — a member of an unofficial WWII flyer's club, each member of which carried a one dollar bill (a one pound note in England) autographed by fellow short snorters. Any member being unable to show the bill upon request of a fellow member, had to forfeit a comparable bill or note to each short snorter present. — Jordanoff, *Jordanoff's Illustrated Aviation Dictionary*. A variation on this has it as a custom that a pilot show a bill of the currency of the country in which they meet, or a piece of paper currency representing every country he had flown into, the defaulter being obliged to buy a round.

The most authoritative story about the origin of this association calls it an "unofficial fraternity relating to accomplishment of trans-oceanic flight. Many concepts of origin and eligibility to membership. Members exchange signatures on paper money attached together, usually from various countries in which they have served." —*AAF: The Official Guide to the Army Air Forces.*

show-and-tell — flight attendants' jocular way of describing the preflight safety/emergency drill.

the **showline** — at air shows, the line of performance, parallel to the crowd, running along the runway on which takeoffs and landings occur and extending vertically as high as the eye can see. "And....he's right on the showline for that **Immelman**!......"

shrimp boat — a piece of cardboard on which the flight plan, destination, estimated time of arrival, etc., are entered for easy visibility in the air traffic control room, and which gets passed along from controller to controller as the flight makes progress in its trip, like a "shrimp boat." Could also be applied to data notes passed around the cockpit. Now known officially as "flight progression strips."

side arm — See **side stick.**

sideslip — to **crab** or fly in a crosswind.

sidestep — air traffic control term used to clear a landing aircraft: When approaching one of two parallel runways, to sidestep over, and land on the other (parallel) runway.

side stick — a **joy stick** on one side of the cockpit, rather than the traditional central location, such as on the Airbus. Also known as **side arm**.

Sientese, Agarase y Matase — Spanish jocular explanation meaning "Sit down, hold on, and die," said to be the meaning of the letters SAM (acronym for the Colombian airline Sociedad Aeronautica de Medellin). "Taking to the skies can be adventure in absurdity," *Toronto Globe and Mail*, July 4, 1994).

Sierra Hotel — phonic code for **shit hot**.

sigmet — significant meteorological signal, notice or alert.

silk bloomers — those who have bailed out.

silk canopy — slang term for parachute, **jumpsack**.

silver bullet — "In the jargon of the trade, a silver bullet was a deadly secret weapon kept under tight wraps until it was ready to be used to take out an enemy in a **Delta Force** covert **surgical strike**." — Rich, *Skunk Works*.

single packet — US Navy expression for an aircraft capable of both search and strike tactics in anti-submarine warfare.

Single Sky for Europe — term applied to the concept of improved, modernized air traffic control over an increasingly congested, yet increasingly unified Europe. **Open skies**. Also known as **EATCHIP**.

sisters — "colours of the day," cartridges used by the British and Allied forces for identification; colours changed daily. Germans tried to but did not succeed in imitating the colours. —Doug Sample CD, Canadian Branch, Yorkshire Air Museum.

six o'clock — usually in plane talk, refers to the position above and/or behind you, from the compass bearing, or more specifically, the compass bearing as stated as if it were a clock face. If you're in combat and an enemy is there, he has the "six o'clock advantage." See **check six, watch your six**.

sixty minute rule — formerly, a rule requiring twin-engine airliners flying extended operations over water (**ETOPS**) to "fly a route that would leave them no more than 60 minutes from an airport at single engine flying speed." During the '80s, increased reliability of "modern turbofans" permitted relaxing of this rule to 180 minutes. El Al's Boeing 767 operated the first commercial **ETOPS** flight in March 1984.

skid — applying opposite rudder and aileron causing the aircraft to slide away from the turn. Compare **slip**.

skip bombing — technique used against dams, ships, etc. where bomb skips across water. The **skip bomb** was also known as the **bouncing bomb**.

to get the **skinny** — to find out the rules, specs, procedures, for a mission or aircraft.

skirts — wheel fairings. Also known as **spats, tin drawers, trousers, pants, wheel pants**.

skunk works — Lockheed Advanced Development Department. According to Jay Miller, "the origin of the name ... is generally attributed to Lockheed engineer Irv Culver, a serious follower of Al Capp's *Lil' Abner* comic strip. In the strip, 'Skonk Works' referred to an illegal still where 'kickapoo joy juice' was brewed. Culver answered the phone one day with 'Skonk Works. Inside man Culver speaking.' The name stuck. After some litigation, Lockheed changed the spelling from Skonk to Skunk and copyrighted it." —*The Skunk Works: The First Fifty Years*. Ultimately, any engineering division devoted to development, brainstorming, etc., was called a "skunk works." But McDonnell Douglas calls its group "Phantom Works."

skyboard — "surfboard-like object used in **sky surfing**." —John and Adele Algeo, "Among the New Words," *American Speech*, Spring 1994.

skycap — an airport porter.

sky cover — the amount of sky obscured by clouds or other phenomena.

skymarking — Wanganui parachute flares used when visibility was very poor, in WWII bombing raids. Bombers would aim at slowly descending flares.

sky pilot — nickname for a chaplain or padre. Also known as "devil dodger" and "God botherer".

skying out — among hang glider pilots, flying so high the pilot may reach 18,000 feet, by FAA rules, the highest permissible altitude in uncontrolled air space. See also **specking out**.

skyjack — Jocularly, "device for changing tires in flight." Actually, aerial hijack of an airplane, usually by passenger(s) wishing to go to a destination other than that originally intended. One famous early example is the WWII episode wherein RAF POWs took over an Italian Cant Airone seaplane and flew it to Malta. See **Fidel**.

sky jockey — "fighter pilot (UK source)" — From John and Adele

Algeo, "Among the New Words: Gulf War Glossary," *American Speech*, Winter 1991. See also **Sabre jock, jock, driver, puke.**

skymarkers — flares dropped by Pathfinders which illuminated the sky above the target, during WWII attacks. —Dunmore and Carter, *Reap the Whirlwind.*

sky shouting — a method of air-ground communications whereby an aircraft flies low and slow over persons on the ground and conveys information to them verbally, often using a bullhorn or other audio-amplification device. See **Hollering Huey.**

sky surfing — "jumping from an airplane with a **skyboard** strapped to one's feet, on which one rides wind gusts before activating a parachute....If you think bungee jumping and rollerblading are dangerous...." —John and Adele Algeo, "Among the New Words," *American Speech*, Spring 1994.

slab control — refers to degree of control pilot has on flight surfaces, or "slabs" - ailerons, elevators, rudder, etc.

slam dunk — "landing technique that allows the plane to stay above traffic until the last minute, at which point it quickly drops to land." — Dickson, *Slang!*

SLAR — Side Looking Radar, a detection device usually mounted on reconnaissance aircraft to scan large areas.

slats — movable control surfaces on the wings, as on the WWII Short Takeoff and Landing Stinson L-1 Vigilant liaison plane, which had "full-width, leading-edge automatic slats" which would "claw us into the air." —"World War II Vigilant: Scarce, Stable and STOL!" *Private Pilot* (September 1984).

sleep debt — accumulated sleep shortage, a common ailment among flight crews, especially those assigned to long hauls.

slick — a stripped-down combat aircraft used for rescue, reconnaissance, or transport. —Mersky and Polmar, *The Naval Air War in Viet Nam.*

sling wing — slang name for a rotary wing aircraft (helicopter). —Heath, *CW2*. **Chopper, copter, egg beater, whirlybird.**

slip — 1. (v.) applying the same rudder and aileron, causing the aircraft to slide into the turn. Compare **skid.**

2. (v.) among parachute jumpers, pulling parachute risers to cause lateral movement of a parachute. —Pat Dudley.

2. (n.) Jocularly, "apparel worn by some pilots."

slot — place within the **ATC** system, "one airplane, one slot" applies to gate, runway, departure, enroute, approach, landing, runway, gate (reserved times).

slot time — time "reserved" with the control towers for an aircraft to take off, land, etc., subject to minor variations caused subsequently by changes in winds, etc.

slow roll — aerobatic maneuover in which the airplane rotates 360 degrees through its axis and returns to normal flight.

smart skin — active anti-radar transmitters embedded in an aircraft's skin to cancel out search radar beams, thus contributing to stealth. —Coonts.

Smoking Crater Airport — same as **smoking hole**.

smoking hole — in the ground, a term for the result of a bad crash, from "Straight and Level," a column in *Flight International*, by "Uncle Roger." See **auger in, buy the farm**.

smoking materials — Paul Dickson has an eye for euphemism: "cigarettes, in the strange cant of the public address announcement. Smoking materials are never put out, they are extinguished." — Dickson, *Slang!*

snag — general Canadian term for any problem with an aircraft, as distinguished from the U.S. **squawk**.

snakeye — "a snake-eye bomb, a bomb with fins that extend to retard the drop of the bomb. A snakeye allows the dropping aircraft to be very low over the target and yet avoid the explosion of the dropped bomb." Vietnam. —Col Bob Stoffey, *Cleared Hot!*

snap roll — aerobatic maneuover started about 30 knots above stalling speed. Bring the stick full back, simultaneously using full rudder and aileron in the desired direction. The nose snaps up and the roll takes place with great suddeness. See also **one "G" roll, flick roll**, and for a somewhat different definition see **roll**.

snooper airplanes — surveillance aircraft. "I always felt that we were more closely monitored on these flights than we probably realized by special RC-135 snooper airplanes packed with powerful electronics." —Lt. Colonel Buz Carpenter, quoted in Rich, *Skunk Works*.

snort — among "stewardesses," "oxygen," according to John Davis, *Buzzwords*.

Snowbirds — Canadian aerobatic team. See **Red Arrows, Blue Angels, Thunderbirds.**

Snowdrop — RAF policeman (WWII). —Doug Sample CD, Canadian Branch, Yorkshire Air Museum. Also known as "snoops."

snow fence — a type of slotted flap, such as on the Japanese Kawasaki Ki-48 Lily of WWII.

snow plough — (Brit.) a contraption of cables running from nose to wingtips, used to cut barrage balloon cables. WWII.

snuffies — paratroopers in training, before their sixth jump. See also **cherry jumpers, prop blast party.** — Pat Dudley.

S. O. B.s — "souls on board," i. e. passengers and crew aboard an aircraft. See also **PAX, talking ballast, self-loading cargo.**

socked in — grounded by weather, especially fog, or an airport thus closed. See **fogged in.**

S. O. E. — Special Operations Executive, a task force group to drop enemy agents and supplies for Resistance forces over enemy territory. —Doug Sample CD, Canadian Branch, Yorkshire Air Museum.

solid overcoat — pilots' slang for "solid overcast," a layer of cloud you can't see through.

2. Another use of the term "solid overcast" was "a large number of aircraft," as in the 1949 movie "12 O'clock High," in which Gregory Peck predicts, in the dark days of WWII, that "some day soon, someone will look up and see a solid overcast of American bombers, headed for the heart of Germany, to get him where he lives."

solo dunk — a naval ritual in which a student pilot, upon completion of his or her first solo flight is tossed into the nearest body of water, usually the ocean. See **tear his shirt.**

sonic boom — a loud bang, actually two distinct bangs, audible on the ground for miles around, caused when an aircraft flies through the **sound barrier.**

sortie — "A sortie equals one a/c making one operational flight. Therefore ten sorties can mean ten a/c flew one mission or one a/c flew ten missions." Originally from the French for "exit" or "go out." —Deighton, *Fighter.*

sound barrier — not an absolute barrier, though it seemed so for a time. When aircraft began to be able to travel as fast as the speed of sound (approx. 760 mph, Mach 1), they created a shock wave, the turbulence of which caused severe testing of pilot and equipment. Chuck Yeager broke through the sound barrier in 1947. Ernst Mach, a scientist/engineer, studied the phenomenon and lent his name to the measurement of speed in these units: Mach 1 is the speed of sound, Mach 2 twice, etc. See **thermal thicket, biological barrier, baby barrier, sonic boom**.

soup — thick weather. Short for **pea soup, muck**.

sour — code meaning a tanker is dry, not available to transfer in inflight refueling. See **sweet**. —Coonts, *Flight of the Intruder*.

Space Available — a basis for granting a seat, usually to an employee of the airline, to permit travel without a confirmed seat. —American Airlines training materials.

Spam Can — any mass-produced light aircraft.

spar — the main wing structural support member.

spark chaser — an electronics technician.

sparkle — "target marking by a gunship using incendiary rounds." —Canadian Forces Aircom Regulation Glossary.

spats — wheel fairings. See **pants, tin drawers, wheel pants, skirts, plus fours, trousers**.

spatmobile — "the Toronto vehicle used by Airport's Special Assistance Team to transport passengers with special needs within the airport, usually in the domestic area." —Mary Manni, "Say What? A guide to airline slang," Air Canada's *Horizons* magazine, April 27, 1994.

specking out — among hang glider pilots, flying so high you appear to those on the ground as a speck in the sky. See **skying out**.

Speedbird — Radio call sign for British Airways flights. Originated with British Overseas Airways Corporation (BOAC) who named their aircraft Speedbird.

speedbrakes — same as **airbrakes**, flaps for slowing down, especially in a dive or a steep descent. Also **dive brakes**.

spin — the spiralling, falling action of an aircraft, sometimes out of control, often the result of a **stall**. There are the following various types, according to Williams, *Casey Jones Cyclopedia of Aviation Terms*: flat spin - rotation around the vertical axis with the nose down so that the longitudinal axis is at an angle of less than 45 degrees with the horizon; inverted spin - a normal spin executed with the airplane in inverted flight position so that the pilot is on the outside of the turn; abnormal spin

- a spin induced by extreme operation of the controls or by other abnormal conditions calling for equally excessive application of controls in the opposite direction.

spin up, spool up — terms currently in use with metaphorical applications in the business world which seem to have come from the world of computers; in fact, they are phrases from old engineers' and mechanics' talk concerning turbines.

Spitfire snobbery — term used by RAF WWII Hurricane pilots to describe the prevailing mind-set that the Spitfire was the best fighter aircraft of the time, a view widely held in England and even among the Luftwaffe. "This was acknowledged by the disproportionate number of Messerschmitt fighter pilots who, on becoming POWs, insisted they had been shot down by Spitfires." —Deighton, *Fighter*.

spit out — "an unintentional exit from the engagement; normally implies a request for vectors back to the fight or to the nearest threat." —Canadian Forces Aircom Regulation Glossary.

spitting out the wire — U.S. Navy term for disengaging a landed airplane from the arresting wire on an aircraft carrier. "Once a plane was in a wire and stopped, it would be pulled backward slowly so the wire would drop out of the hook." —Wilson.

splat — result of **augering in**. See also **prang**.

split-arse turn — split-S turn.

splits — descending, turning manoeuvre used to escape in air combat. —from Paul Turk.

split tail — see **butterfly tail, v-tail**.

split ticket — see **back-to-back**.

spoofing — "informative that voice deception is being employed." —Canadian Forces Aircom Regulation Glossary.

sponson — a light air-filled structure or a winglike part protruding from the hull of a **seaplane** or **flying boat** to steady it on the water.

spot-on — a navigator's favorite term, meaning arrival exactly on or over a navigational **fix**.

sprog — apprentice pilot — "sprog pilot," "training sprogs to go on to F-86s" —Gord Squires, "Joy Riding in an F-86", *Aircraft* (April-May 1990). Doug Sample identifies it as term for a "newly commissioned officer, pilot officer rank, and also in use for a brand new uniform, WWII.

spun in — any bad mistake, also, crashed into the ground. WWII. See **auger in, buy the farm**.

sputter 'n' stutter — "stall," in "stewardess" slang, according to John Davis, *Buzzwords*.

square circuit — manoeuver performed by experienced de Havilland Moth pilots on windy days: "They'd get a bunch of us guys out holding the wingtips of a Moth and they'd take off, full throttle, and climb up, and then fly *backwards* across the field, close the throttle to descend, and then pour on the power to come in. And we'd all be there to grab them as they touched down." Canada, WWII. — Williams, *The Plan*. (The point of the phrase is that the craft would go straight up, make a box by flying backwards, then come straight down).

squadron — a number of **flights**, as a term of quantity and organization of an air force. See also **element, wing**.

squawk — 1) typical generalized term for any problem with an aircraft, in the U.S., as distinguished from the typical Canadian term, **snag**. 2) "The radio transmission of the radar transponder onboard an aircraft; also the Air Traffic Control instruction to the pilot to set one of 4096 possible codes to identify the aircraft on controller radio." —Aircraft Owners and Pilots Association "Aviation-to-English Dictionary", a four-page mainly technical word-list. In "squawk 3 normal," a directive to turn on your identifying transponder (the **parrot**) which emits a signal enabling the other aircraft or control tower to identify the aircraft as friend or foe (**IFF**). See also **strangle the parrot**.

squirted off — launched by catapult. See **cat shot**.

stablerons — horizontal subplanes on nose or tail, for stability.

stacked up — term applied to aircraft waiting in a holding pattern for their turn to land. Such aircraft are said to be "in the stack."

stag's antlers — *hirschgewei*, radar antennae of WWII German design protruding from the nose of an aircraft, usually a Junkers Ju-88. See **morning star**.

staggerwing — in biplanes, a design in which the lower wing is positioned forward of the upper wing, a planform made famous in the Beech 17 Staggerwing, an instance of "negative stagger."

Stainless Steel Airline — one of the few, some say the only, nickname for American Airlines. The name refers to the aluminum finish of the aircraft, and implies invulnerability and reliability.

stall — loss of flying speed and lift because the angle of attack exceeds the angle maximum lift. The types: advanced stall - a flight manoeuver in which the airplane is fully stalled yet with sufficient control remaining so that it can again be nosed down to regain flying speed: whip stall - a complete stall in which the nose of an airplane whips suddenly downward, the plane usually sliding backward on its tail at the same time. — Williams, *Casey Jones Cyclopedia of Aviation Terms*. When you lose **lift**. Jocularly, the place where the airplane is kept.

stand-down — all flying cancelled (WWII). Usually because of d**uff** weather over target area.

stars-and-bars — the U. S. military insignia painted on fuselage and wings of U. S. military aircraft. See **roundel**.

start the lamp swinging — to talk about flying, perhaps connoting some stretching of the truth. See **hangar flying**.

stealth — term applied to any modern-generation aircraft that produces a low visual, radar, or infrared signature, by its shape, and a surface that absorbs and disperses radar signals. In application, not one F-117 Stealth Bomber was hit by enemy fire during the Gulf War.

steam gauges — the "instrument cluster in older jet aircraft," before the **glass cockpit**, according to Barry Lopez.

St. Elmo's fire — "Defined as a corposant (holy body from the Latin corpus sanctum), St. Elmo's fire has long been known to sailors who may witness this electrical brush discharge appearing as a mysterious glow around the masthead. It is also a familiar sight to fliers but ... it would occur in cirrus cloud or even on the ground, around the engine, in very moist conditions..." — Barlay, *The Final Call*.

step — In the early development of seaplanes, Glenn Curtiss introduced a "step" or "break" in the bottom surface of his pontoons. "This reduced the suction of the water, which had kept his early aircraft and those of many of his contemporaries, firmly glued to the water." — Allward, *An Illustrated History of Seaplanes and Flying Boats*.

sterile cockpit — "the time period when no communication is allowed between flight attendants and the cockpit except in an emergency situation." —American Airlines training materials. During take-off and landing, for example.

stew, stewardess — old, politically incorrect, never-heard-any-more term for "flight attendant."

stew zoo — "an apartment where a lot of stewardesses live, or a hotel where they lay over." — Dickson, *Slang*.

stick — 1. the prop. 2. the **joystick** (see also **yoke**). Arguably, a **dead stick landing** is one with the prop not turning, for which see *What's the Gen*, by HW: "dead stick: a propeller stopped in the air." Murray, in "The Language of Naval Fighter Pilots," reports that this word also means "pilot," in addition to "control lever." 3. a group of paratroopers all jumping out of the same side of an aircraft. —Pat Dudley.

stick and rudder — the basic flight controls in any conventional aircraft. The phrase has come to mean "a simple, uncomplicated airplane," or "a relatively simple flying situation." "Stick and rudder ability" is now sometimes called "cockpit management skills," with the coming of the **glass cockpit**. —Barry Lopez.

stick-and-string — slang for early airframe construction using wood covered with fabric. See also **tube-and-rag**.

stick bombing — method of bombing where bombs are released one after the other in hopes of 'walking across' the target. Also **line bombing**.

stick pusher — anti-stall device which puts the nose down when the aircraft nears stall pitch.

stick shaker — an anti-stall warning device that actually shakes the control stick when the aircraft approaches a stall attitude.

stick thermal — in glider talk, when a pilot applies back-pressure to the control stick and causes the glider to climb, he is said to be using the **stick thermal**. Not to be misinterpreted as a **thermal**.

stimulator — nickname for the flight simulator, applied by an Air Creebec Dash 8 pilot, because "you have to deal with all kinds of failures and emergencies which you can't do in a real airplane....like fires."

sting, stinger — 1. slang name given the long thin boom extending from the tail of modern anti-submarine warfare aircraft, housing the **MAD**, the magnetic anomaly detector equipment.

2. machine gun bullet, Royal Flying Corps, WWI.

sting in the tail — term applied to rear mounted jet engines such as on the Boeing 727, Douglas DC-9 and BAC 1-11.

STOL — Short TakeOff and Landing Aircraft, in which these procedures may be accomplished at 100 knots or less of speed.

stooge — stand-in, replacement (WWII military aviation). —Doug Sample CD, Canadian Branch, Yorkshire Air Museum.

stooging about — patrolling an area (WWII military aviation).

stop and go — a landing in which the aircraft comes to a full stop, then takes off without having returned to the end of the runway. **Touch and go** is similar, but without the full stop.

stopway — a continuation of the runway strong enough to support an aircraft in case of an aborted takeoff or an overshoot. Also called an **overrun**.

2. British term for the rectangular area near the end of the runway used to "hold" aircraft awaiting takeoff clearance.

Stormscope — registered trademark of B.F. Goodrich Inc. for its lightning strike and weather radar. Often used generically for any weather radar.

strangle the parrot — turn off your transponder. See **parrot**. "Strangle" means "turn off" any equipment indicated. —Canadian Forces Aircom Regulation Glossary.

strategic bombing — air war technique first tested, with mixed results, in WWII, until the A-bomb proved its worth. (Bombing cities to destroy morale usually stiffened it, on both sides). After that, Korea and Viet Nam saw its failure; the Gulf War of 1991 seemed to represent its triumph, as the Allies used it to destroy Hussein's ability to fight back, destroying communications links, supply routes, etc. —"Can Bombing Win A War," Nova, U. S. Public Television, Jan. 19, 1993. See also **Bomber Harris**.

stream — the long string of bombers used on night and day operations by Bomber Command in WWII. All aircraft had specific heights to fly at and a time to arrive over the aiming point of a designated target. In this way a force of 800 or 1,000 bombers could be concentrated in the dark to pass over the aiming point in 30 to 45 minutes. —Douglas Sample.

stream buzzer — electronic countermeasure activity.

stressed skin — fuselage construction in which the skin serves as an integral load bearer.

strike package — "Combination of bombers, electronic countermeasure aircraft, and escort fighter." — From John and Adele Algeo, "Among the New Words: Gulf War Glossary," *American Speech*, Winter 1991. An early example might be the Battle of Taranto, resulting in the sinking of the German battleship Graf Spee.

student pilot — "a lovable person with no money." — Baxter, *More Bax Seats*.

stuffed cloud — a cloud full of mountain. **Cumulogranite.**

stuka — usually applied to the Junkers Ju-87 dive bomber, but generic during and after WWII to mean any dive bomber. —Deighton, *Fighter*.

on the **stump** — number one in the lineup waiting to take position to take off.

stunt, stunting — term meaning **aerobatics**, dating from the **barnstorming** days. Practitioners are known as 'stunt pilots.' Joyce Spring, in *Daring Lady Pilots*, about the early achievements of women pilots in Canada, lists such specific stunts as "bunts," "the falling leaf," "the wing ding," "inverted spins (starts on its back)," and "Cuban eights," a series of loops and rolls including **flick (snap) rolls, wing walking.**

going **stupid** — when a "smart bomb" or controlled missile runs amok, usually because of a technological breakdown. —U. S. News & World Report, *Triumph without Victory*, 1992.

sty in the eye — among "stewardesses," "poor visibility," according to John Davis, *Buzzwords*.

sucker hole — slang for what looks like a safe path through thunderstorms or cloud layers but can actually lead to areas of strong turbulence or solid clouds.

suicide door — likely first put on some cars, this is a door that opens from the front, pivoting on the hinges nearer the back of the vehicle. The Republic Seabee had them, and in some cases it was said that if the doors ever came open in flight, they'd be torn off and pieces would go into the props, engines, or empennage, causing terrible havoc.

Sunday driver — flies only on weekends. See **weekend birdman.**

supercharger — intake turbine device boosting air flow to the carburetor or fuel injection system. Well known on cars, it came from airplanes first, where the thin air of altitude flying made it useful, and increased performance besides. See also **turbo supercharging.** Jocularly, a pilot with a wallet full of credit cards.

supercritical wing — a wing reduced in size and weight, to provide minimal lift and stability, for which compensation must be made in fuselage and empennage design. See **Coke bottle design, Marilyn Monroe design, area rule.**

super jumbo — projected next-generation jet airliners (in 1994) seating up to 800 **PAX.**

surgical strike — though this phrase is also used in military operations on land and sea as well as in the air, the primacy of air-power in recent years for such surprise, highly effective, massive and precise attacks makes it an air word.

swan, swanned, swanning — RAF WWII slang for flying leisurely, lazily. **Boring holes, punching holes, honking about.**

sweater — among "stewardesses," a slang term, along with **faucet,** for a nervous flyer, according to John Davis, *Buzzwords.*

sweet — code meaning a tanker is ready to transfer fuel in an inflight refueling. —Coonts, *Flight of the Intruder.* See also **sour.**

swept wing — important innovation in design for the first jets. A wing swept back rather than perpendicular (at right angles) to the fuselage creates less drag, aerodynamically causing the air flow to behave as if the wing is much thinner. See **variable sweep.**

swinging the compass — calibrating the compass, usually by comparing onboard compass readings with those on a large **compass rose** painted on the ramp.

swinging the prop — starting an aircraft by manually turning the propeller.

swing-wing — see **variable sweep.**

Swordfish leg — condition developed by Fairey Swordfish pilots during WWII, who had to keep constant pressure on the right side of the rudder bar to counteract torque.

synchronizer — WWI device that governed the rate of fire of machine guns through the propeller arc, harmonizing the fire so that prop blades would not be struck. See **interrupter.**

T-prop — short for turboprop.

tab effect — bent or torn control surfaces which upset the trim of the aircraft.

TAC — Tactical Air Command (USAF).

tac air — "Tactical air power, Air Force fighters or Army attack planes supporting other operations." — From John and Adele

Algeo, "Among the New Words: Gulf War Glossary," *American Speech* (Winter 1991).

Taco Airline — nickname of TACA, Transportes Aerovias Centro America. Also known derisively as Take A Chance Airline.

Tac-R, tac recce — tactical reconnaissance.

Tail-End-Charlie — the rear gunner (British and Canadian use). —from Doug Sample CD, Yorkshire Air Museum, Canadian Branch. (Sample was a Tail-End-Charlie with a Canadian bomber squadron, WWII).

2. last plane in a formation. "Adding a 'tail-end' Charlie to weave from side to side at the rear of the formation, and thus protecting the tail, had resulted in the loss of too many tail-end Charlies. By the middle of July 1940, weavers were seldom seen." —Deighton, *Fighter.*

taildragger — aircraft with "conventional" landing gear, that is, two main wheels and a tail wheel. According to Gordon Baxter in *More Bax Seats*, "a passenger weighing over 250 lbs." Also, a pilot who lost a bout with the bottle last night. See **checking the tailwheel.**

tail feathers — the **empennage**—rudder and horizontal stabilizer.

tail heavy — aircraft tends to fly in a nose-high attitude due to improper loading or trim. Opposite of **nose heavy.**

tailhook — "arresting hook" on carrier-based airplanes. Gave its name to the Tailhook Association, an "old boys' active-duty and retired US naval aviators" association which between 1986 and 1991 subjected females, including qualified pilots in the service, to a "gantlet," an ambush in a hotel hallway in which members would paw, molest, grab breasts, try to remove panties, etc., according to accusations made by US Navy Lt. Paula Coughlin (Montréal *Gazette*, Sunday, June 28, 1992). The resulting scandal was an embarrassment for the US Navy and caused some disciplinary actions, including dismissal.

tailing in — pulling a seaplane ashore tail first. Beaching tail first.

tailslide — aerobatic manoeuver in which the airplane slides downward, tail first.

tailstand — to achieve zero velocity, balancing on your exhaust. The MiG 29 is one of the few that can do it, as demonstrated at the Paris Air Show in 1989. See also **cobra, vertical reverse.**

take it low and slow and throttle back in the turns — see **airline pilots' oral rules**.

talking ballast — crew slang for "passengers," **PAX**, **self-loading cargo**, **S.O.B.s**.

Talking Bomb — nickname of Squadron Leader Richardson, RAF, leading expert of bomb sights, bombs, and bombing techniques. — Brickhill, *The Dam Busters*.

tallboy — 12,000 pound bomb, in WWII. See also **blockbuster, cookie, doodlebug, Grand Slam, earth quake bomb**. —Doug Sample CD, Canadian Branch, Yorkshire Air Museum.

Tally-Ho! — what a pilot responds to the air traffic controller when he actually spots a nearby airplane he has been told to watch for, to signal his recognition. See also **no joy**. RAF term from WWII, indicating "enemy sighted." Phrase borrowed from fox-hunting.

Tame Boar — German fighter operations that infiltrated the bomber stream to find and attack bombers on the way to and from the target. WWII. See also **Wild Boar**. —Doug Sample CD, Canadian Branch, Yorkshire Air Museum.

tandem — aircraft with pilots positioned one behind the other.

Tango Uniform — phonic code designation for "tits up," "broken." See **Uniform Sierra**. **Tits up** can also mean "shot down." See **Bravo Delta**.

tankers — aircraft adapted to carry fuel as cargo. In-flight refuelers. Also, in car racing, a name for certain **lakesters**, early drag racing machines in which the body was made out of reused aircraft belly fuel tanks. Also known as **belly tankers**.

2. — usual pilots' word for **water bomber**.

TAP — Transportes Aereo Portugales. Also, derisively, Take Another Plane.

taps — Brit., the various switches, dials, and levers in the cockpit.

tapes — in WWII military aviation, non-commissioned officer's stripes. —Doug Sample.

 tarmac — paved surface at an airport. Also **apron, ramp, pan**.

TBO — Time Between Overhauls.

teardrop canopy — a clear Plexiglas teardrop-shaped canopy cockpit enclosure such as on the P-51 Mustang or the F-86 Sabre.

teardrop landing — a landing involving a manoeuvre that changes direction 180 degrees, shaped like a teardrop. Compare **90-270**.

tear his shirt — a ritual in pilot training. After first solo flight, the older pilots tear a piece off the new pilot's shirt and tack it to the bulletin board, with the date. Navy pilots were tossed into the sea. See **solo dunk**.

tease trip — among flight attendants, "a flight to a desirable location on a schedule that leaves little or no time to spend there. 'We landed in Bali Tuesday night and had to fly back to the States the next morning. It was the biggest **tease trip**.'" —L.A. Speak, Los Angeles *Times* magazine, March 20, 1994.

Tee Emm — "Training Memoranda," a series of pilot training updates issued by British Air Ministry during WWII, made famous by Raff and Anthony Armstrong through their cartoon character Pilot Officer Prune.

Teeny Weenie Airline — an early nickname for Trans World Airlines, later also known as **Try Walking Across, The Worst Airline**. See **TWA**.

telecopter — pun on "helicopter," a live, television-camera equipped news or traffic **whirlybird**. See **eye in the sky**.

tenpins — "unattended children," among "stewardesses," according to John Davis, *Buzzwords*. See also **Etna, poker, claw, faucet, sweater**.

test bed — any aircraft used to test new equipment, usually engines or avionics. Often referred to as a "flying test bed." The original Boeing 747 was so used to test the new engines for the 777, with one engine mounted in place of one of the four 747 engines.

tetrahedron — an airport device used to indicate wind direction, shaped like an airplane, visible from the air. See **wind sock**.

thatch weave — WWII fighter formation. See **fluid four**.

thermal — student pilot's description of a container for hot coffee; in fact, "an updraft of air resulting from sun-heated terrain, which causes aircraft, especially gliders, to climb."

thermal thicket — technical bugbear next in line after the **sound barrier**. At about Mach 3, air friction heats the skin of an aircraft to the point above where aluminum weakens, about 500 - 600 degrees F. See also **baby barrier, biology barrier, sound barrier**.

thermalling — in soaring, using thermal updrafts to gain altitude.

thimble nose — protruding radome, thimble-shaped, as on the de Havilland Sea Hornet.

three greens — three green lights on the instrument panel indicating "wheels down and locked."

three-holer — three-engine aircraft such as B727, L1011, DC 10, Tu 154, Yak 40, Yak 42, Trident. In early days, a biplane with three cockpit openings. Also, an aircraft with three lavatories, according to John Wheeler and Jack Gamble of Boeing. See **one-holer, two-holer**.

Three Musketeers — US Army Air Service aerobatic team flying biplane pursuits during the 1920s.

three-pointer, three-point landing — a perfect touchdown, with all three wheels meeting the ground at the same time. See **on the numbers**. A Chinese three-point landing is a crash landing, "the three points being the propeller hub, one wheel, and one wingtip." —Berrey and Van den Bark, *The American Thesaurus of Slang*. Also known as a **daisy cutter**.

three-surface airplanes — aircraft that have **wings, empennage** and **canards** as **control surfaces**.

threshold — the end of the runway, or earliest possible spot on the ground at which touchdown is safe and desirable.

through the gates — to accelerate as rapidly as possible, so named because the throttle is a hand-operated stick which can only be advanced through a series of off-set gates, at each of which the pilot must move the control sideways before advancing it further. "To get maximum power, literally, by placing the throttle lever through a gate on a quadrant." —W. H. *What's the Gen?* See **firewall it**.

thumping — "game played by two fighter pilots—and condemned by their superiors—in which one plane sneaks up on the other, flying below it, then accelerates sharply and loops up directly in front of the other plane's nose. The effect of "thumping" is not only to scare the other pilot but to disturb the air currents around the other plane's wings, thus making manoeuvreing extremely difficult." — Murray, "..Navy Fighter Pilots."

Thunderbirds — The USAF aerobatic team. See **Blue Angels, Snowbirds, Red Arrows**.

tick-a-lock — a flight attendant who stays in her room on layovers.

ticket — license, more fully, "private ticket," "commercial ticket," etc..

Tiger Force — RAF bomber squadrons assigned to bomb Japan after VE Day.

TIGHAR — The International Group for Historic Aircraft Recovery, a Delaware-based organization interested in recovering and restoring downed aircraft of historical significance.

tilt-rotor — "Airplane that takes off and lands like a helicopter but that tilts the rotor blades to fly like a fixed-wing airplane." — From John and Adele Algeo, "Among the New Words: Gulf War Glossary," *American Speech* (Winter 1991).

tin bird — airplane. —Berrey and Van den Bark, *The American Thesaurus of Slang*.

tin can — WWII and later usage for a destroyer, in pilots' slang. — Searls, *The Crowded Sky*.

Tin Can Air — see **Can Air.**

tin drawers — wheel cowlings. —Berrey and Van den Bark, *The American Thesaurus of Slang*. Also **plus fours, pants.**

tin fish — naval aviator's slang for torpedoes.

Tinkertoy Airlines — see **Treetop.**

tin kickers — "aircraft accident investigators." Also known as **jigsaw mullahs.** See **Go-team.**

tin tie — ("zinnbinden") German slang for the Knight's Cross, which was worn on a ribbon, at the neck. Highest award in Hitler's Germany. —Deighton, *Fighter*.

Tip-n-Tell Label — "a device that, when affixed to a shipment, indicates whether the shipment has been tilted beyond a 40 degree angle." —American Airlines training materials. See also **shock watch.**

T I T — Turbine Inlet Temperature.

tit — bomb release button, RCAF WWII. —Williams, *The Plan*. Also applied to any push-button switch.

tits up — shot down, also broken or unserviceable. —Jess Finney, USAAC, San Diego. See **Tango Uniform, Bravo Delta.**

TLAR bombsight — on fighters, "supersecret" item, explained as meaning "That Looks About Right." — Stevens, *"There I was....."*

toboggan — "the word announced by a receiver pilot as an indication to the tanker pilot to commence a predetermined rate of descent." —Canadian Forces Aircom Regulation Glossary.

toe tag — among "stewardesses," a boarding pass, according to John Davis, *Buzzwords*.

toilet-seat flying — according to Doug Chivers, Air Canada DC-9 pilot, what you do a lot in that aircraft—short runs, "up and down, up and down," such as runs to Newark, Regina, Winnipeg, Fredericton. —Mitchell.

toga button — "take off and go around" switch for auto pilot.

Tokyo tanks — WWII term for auxiliary gas tanks. —*AAF: The Official Guide to the Army Air Forces.*

Tombstone Agency — irreverent name for the U.S. Federal Aviation Agency (FAA) because "they only act when there's been a fatality."

tombstone engineering — redesigning an aircraft or component to remove a fatal characteristic, usually after the flaw has revealed itself the hard way.

toothpaste — among "stewardesses," "severe injuries caused by a sudden violent depressurization," according to John Davis, *Buzzwords.*

top cover — presence of fighters at a higher altitude to provide protection for lower flying aircraft or ground troops. See **capped**.

top gun — "navy fighter weapons school in San Diego, California that teaches tactics and doctrine, or a student at or graduate of the school." —Murray, "..Navy Fighter Pilots." "Initially called U. S. Navy Postgraduate Course in Fighter Weapons Tactics & Doctrine, the school became known as Top Gun and occupied a hangar on the master jet base at Miramar, CA, northeast of San Diego. The first class convened in March, 1969. Special emphasis was placed on what one pilot called 'old-fashioned **dog-**

fighting with new-fashioned weapons.'" —Mersky and Polmar, *The Naval Air War in Viet Nam.*

top side — in the air (WWII). —Doug Sample CD, Canadian Branch, Yorkshire Air Museum. **Upstairs,** up **Alice Blue Gown.**

torching — the burning of fuel in the exhaust outlet, the result of a too-rich mixture. Not to be confused with **afterburner, reheat.**

tore him off a strip — "rebuked him." —WWII bomber slang. —from Raff and Armstrong, *Plonk's Party.*

torque stall — on waveoff or abort, when full power is reapplied, aircraft tends to pull left (or right) due to sudden prop torque. Aircraft goes into nose-up, turning stall in which even hard opposite rudder is ineffective.

torque swing — tendency of aircraft to fishtail due to high engine torque.

touch and go — practice landing followed by immediate take-off. See **circuits and bumps**. According to Gordon Baxter in *More Bax Seats,* "an airline pilot with a wife at both ends of his route."

touch bottom — crash (WWII military aviation). —Doug Sample.

tour — thirty bomber operations or 120 points (3 points for a heavily

defended target, 2 for one lightly defended), in WWII. Completing a second tour required 60 points.

tracer — ammo projectile (WWII) with base end that glowed. There were both day and night tracer, the night tracer having a dim glow so as not to hinder a gunner's night vision. One type was **de Wilde**.

track — the line the aircraft takes over the ground.

tractors — "puller" (propeller) engines. See **pusher**.

driving the train — leading more than one squadron into combat (WWII military aviation). —Doug Sample.

TRAM — Target Recognition Attack Multisensor, a turreted electro-optical/infrared sensor system coupled to laser-guided weapons for attacking ground targets.

transat — transAtlantic. Opposite is **domestic**.

Trans-Traumatic Airlines — see **Treetop**.

trap — "Landing of a plane on a carrier." — From John and Adele Algeo, "Among the New Words: Gulf War Glossary," *American Speech* (Winter 1991).

trash air — among hang glider flyers, turbulent air. The opposite of **fat air**.

trashcan — see **ashcan**.

trash hauler — freighter, military cargo plane, and pilots thereof. —from Paul Turk.

Treetop, Treetopper Airlines — nickname of Trans-Texas Airways, later Texas International before being absorbed into the Continental system. Also **Tinkertoy Airlines, Trans-Traumatic Airlines**.

tricycle landing gear, carriage, aircraft — aircraft with nose wheel. See **conventional aircraft, taildragger**.

trim — term of rough measure to describe the complex adjustment of control surfaces to keep the airplane flying on course and level or banked, whatever's required. Many light aircraft have the trim control cranks mounted on the overhead, prompting the question, "which way do I turn it?" In most aircraft the answer is "clockwise for nose up."

trim tabs — small control surfaces to do fine tuning on the **trim**.

trim the trees — among "stewardesses," a slang term for "take off," according to John Davis, *Buzzwords*.

triphibian — landing gear or apparatus allowing aircraft to operate from land, water, and ice/snow

—such as the Grumman Albatross.

triple A — Anti Aircraft Artillery. See **flak, Archie, ack-ack**.

troll — "To fly without a predetermined target, searching for targets to bomb." — From John and Adele Algeo, "Among the New Words: Gulf War Glossary," *American Speech*, Winter 1991.

trombone — "the sliding lever to defrost the canopy" on an F-86. —from Gord Squires' "Joy Riding in an F-86", *Aircraft* (April-May 1990).

troops — slang term for mechanics and ground crew during WWI. See **black gang**.

trousers — wheel fairings. Also known as **tin drawers, pants, skirts, spats, wheel pants**.

trucks — of wheels, e.g. 4-6-16- on the undercarriage of an aircraft.

true speed — corrected speed from indicated airspeed and other factors.

Try Walking Across — nickname for TWA. See also **Teeny Weenie Airline**.

T-tail — empennage configuration in which the horizontal plane is at or near the top of the vertical stabilizer. The more common configuration would have the horizontal plane near the middle or at the bottom. Also distinguished from the **V-tail**. Jocularly, "fashion designer influence on light aircraft." — Baxter, *More Bax Seats*.

tube-and-rag — airframe of tubular metal or aluminum construction, covered with fabric. See also **stick-and-string**.

tubs — slang for thrust reversers.

tuck — phenomenon associated with **compressibility**, describing an aircraft's tendency to tuck nose down "when encountering compressibility, an early problem in the Lockheed P-38 Lightning test-flight program. Often stated as 'compressibility tuck.'" —Deighton, *Fighter*.

tufts — small pieces of string attached to flying surfaces to detect variations in airflow, or, as in the McDonnell Douglas MD80, wing ice caused by **cold-soaked fuel**.

tumbleweed — "indicates lack of total situation awareness; no tally, no visual; a request for information." —Canadian Forces Aircom Regulation Glossary.

turbojet — jet airplane, because the engine is a turbine which delivers its power in the form of a jet of hot gases. See **kerosene burner**. In London, January 16, 1930 — "a patent for a new kind of airplane engine was filed today by a junior officer in Britain's Royal Air Force. Flying Officer Frank Whittle, a flying instructor and former fighter and test pilot, believes that aircraft can be powered by what he calls a 'turbojet,' a gas turbine engine which creates a propulsive jet of hot gas." — Gunston, *Chronicle of Aviation*. For a good description of the distinctions between "turbojet," "high bypass jet engine," and "turbofan," along with an explanation of why there are rarely "pure jet engines" on civil aircraft, see Gary W. Salewicz' "Major Developments in Jetliner Technology" in the **ICAO** 50th anniversary book.

turbo prop — aircraft whose propellers are driven by a jet (turbine) engine. See **kerosene burner, jetprop.**

turbo supercharging — has taken over the technology because there is no mechanical connection between the motor and the mechanism which compresses the gas/air mixture. Exhaust gases are cycled through an impeller which does the packing: thus it's an air interface, unlike the early superchargers, which used a set of cams, gears, and valves, cumbersome and unreliable, and productive of enormous additional noise (see **Blower Bentley** in Poteet and Poteet's *Car and Motorcycle Slang).* —from Don Hackett. "It's, like, hooking up a vacuum to the carburetor!"

turkey shoot — "Destruction of easily reached and undefended ground targets by aircraft. (WWII use)" — From John and Adele Algeo, "Among the New Words: Gulf War Glossary," *American Speech* (Winter 1991).

turnaround — "flight that returns a crew or crew member to their home base on the day of departure." — Dickson, *Slang!* Also, to land, unload, reload, and take off for the next destination—"quick turnaround."

turn 'em and burn 'em — "To service aircraft after a mission and get them back on another mission." —From John and Adele Algeo, "Among the New Words: Gulf War Glossary," *American Speech* (Winter 1991).

Tuskegee Experiment — see **Redtails.**

TWA — Trans World Airlines, also known jocularly as The Worst Airline, **Try Walking Across,** (another vaiation: Try Walking Again), **Teeny Weeny Airline.** "I hate TWA coffee, but I love TWA tea."

twelve hours from bottle to throttle — rule of thumb on how long before flying a pilot must abstain from consuming alcohol. See also **eight hours......**

twilight zone — a term used by pilots to describe the phenomenon produced when one descends so low to the runway that the horizon disappears and an abrupt feeling of imbalance and disorientation ensues until a visual reference can once again be established. The famous Rod Serling mystery series from 1959-64 was named with this aviation phrase. See **pucker zone**.

twilights — WAAF (WWII) issue knickers, summer weight and lighter coloured than **blackouts**. —Doug Sample.

twins — twin-engined aircraft. In the world of modern jet aircraft, "according to Boeing, twin-engine operations offer airlines five to nine percent lower operating costs over three and four-engines aircraft." But see **sixty minute rule**. —ICAO 50th anniversary book.

twitch, the — uncontrolled trembling of head and facial nerves (WWII bomber crew) due to long duty on **ops** and exposure to fighter attacks and flak. Usually came on after mission or tour (25-30 missions) completed. After tour, aircrew went on leave followed by six months teaching. Third tour was voluntary.—Douglas Sample CD, Canadian Branch, Yorkshire Air Museum.

two-bomb theory — hypothetical (we all hope) cynical and eccentric theory of self-protection from terrorist bombs on airplanes. As the odds against there being two bombs on one airplane are much more favorable than those of there being one, you carry your own bomb aboard!

two-holer — in open cockpit airplanes, the opening at the top of the cockpit through which the pilot's head protrudes is called a hole. A two-holer is a two-place airplane, with the pilots sitting in tandem. Also called a 'deuce.' See also **one-holer, three-holer**.

two-man sled — "fighter plane that holds two passengers [sic], a pilot and a radar intercept officer." —Murray, "Language of Naval Fighter Pilots."

two-stage amber — WWII RAF synthetic night-flying gear comprising amber cockpit screens and blue glasses to simulate night flying.

two-versus-two hop — "simulated air battle between two student crews and two crews of instructors." —Murray, "Language of Naval Fighter Pilots." See also **one-versus-one hop, two-versus-unknown hop**.

two-versus-unknown hop — "simulated air battle between two student crews and an unknown number of crews of instructors." —Murray, "Language of Naval Fighter Pilots." See also **one-versus-one hop, two-versus-two hop**.

U-bird — popular term for any utility aircraft. See **L-bird**.

ughknowns — what test pilots call the unknown qualities of test aircraft. —Chuck Yeager's autobiography. See also **unk-unks**.

unable — term used in radio communications to mean "unable to comply."

undercarriage, undercart — British term for landing gear.

undercast — it starts out as overcast when you're on the ground and becomes undercast once you're **on top**, in the clear.

under the hood — See **hood work**.

under the weather — below the clouds, often at low altitude. **Scud run**.

unicorn — what you call an aircraft which, during an air-to-air refueling procedure, breaks off and takes the boom with it, as the boom protrudes. Also known as a **pogo stick**.

Uniform Sierra — **unserviceable**, or u/s, in the phonic code. See **Tango Uniform, Bravo Delta**.

unk-unks — aircraft (specifically air*frame*) industry term for "unknown factors." — Logie, *Winging It: The Story of the Canadair Challenger*. According to Karl Sabbagh, some of them are market forces,

costs — e.g. oil, and the availability, site, climate, and altitude of the changing airports which are fashionable travel destinations, over time. (It takes a long time to take an aircraft from idea to reality).

unobtanium — "a substance or a piece of hardware that is desired but not obtainable." (Space talk) — Dickson, *Slang!* This term also occurs in automotive mechanics' talk: see Poteet & Poteet, *Car & Motorcycle Slang*.

unserviceable — broken, often spelled u/s. **Uniform Sierra**.

unstart — sudden loss of power due to a disturbance at the engine air inlet shock cone. At speeds like Mach 3 the shock cone must be positioned exactly on the inlet or air intake will be interrupted. Felt as a sudden jolt. Typical of the Lockheed SR-71 Blackbird. "It occurred when air entering one of the two engines was impeded by the angle of the airplane's pitch or yaw and in only milliseconds decreased its efficiency from 80 percent to 20 percent." —Rich, *Skunk Works*.

unstick — point at which the aircraft leaves the ground or water, **lift-off**.

up and locked — refers to the landing gear, the proper position thereof once the airplane has left the ground. In early retractables, pilots would sometimes forget to retract the wheels and then wonder why their airspeed was so low and their fuel-burn so high. Said of the landing gear, and when a mistake has been made, of the perpetrator's head. "...had my head up my ass....." See **down and locked.**

up gripes — less serious aircraft deficiencies, entered on a **gripe sheet,** than **down gripes.** Repairs to be done as time and opportunity allows. —Coonts, *Flight of the Intruder.*

USAAC — Army Air Corps, American, 1926-41. Previously USAAS, Army Air Service (1918-26). Later called the USAAF, Army Air Force (1941-47), and as of 1947, USAF. The AAC was memorialized by a song, by Airman Robert Crawford:

Off we go, into the wild blue yonder,
Climbing high into the sun;
Here they come, zooming to meet our thunder,
At 'em boys, give 'er the gun!
Down we dive, spouting out flame from under,
Off with one helluva roar,
We live in fame or go down in flame, hey!
Nothing'll stop the Army Air Corps!

USAF — United States Air Force (1947-). See **USAAC.**

Useless Air — nickname for USAir.

U.S. formation — a four-ship flight with the elements staggered and flexible, giving the leader visual control, common in the Pacific theatre during WWII. —Van Heatherly, "Charles Lindbergh's Heroic Role in World War II," in *Fighting Eagles,* ed. Phil Hirsch.

UTA — Air France's Africa connector subsidiary, Union de Transport Aerienne, nicknamed Unlikely To Arrive.

V — code for "velocity," with various quite technical uses. V_1 is a speed arrived at by various calculations based on aircraft, etc., at which the plane should be able to take off, even if it loses an engine; V_2 is a "second-stage calculation" which describes the speed at which aborting the landing is not necessary (even in case of engine loss) after takeoff, with flaps and other gear still out but wheels up and before "cleaning up," or trimming back flaps etc. for minimum drag (see **dirty**). Vs is stall speed; at the opposite end, Vne means "Velocity Never Exceed," or a speed at which G-forces will be so great that there's a danger of wings being torn off. V_{mu} is "minimum unstuck speed," or the speed at which the craft may become airborne. See **Vee formation**.

variometer — an instrument used mainly in gliders which shows slight changes in attitude.

V-Bombers — a series of nuclear bombers operated by the RAF during the 1960s. They included the Avro Vulcan and the Handley-Page Victor.

V-tail — Vee-shaped tail, also called **butterfly tail, split tail**, utilising a **ruddervator** for combined rudder/elevator control surface, such as on the Beech Bonanza.

vapor trail — **contrail**. A white streak across the sky caused by the condensation of engine exhaust vapor at high altitude.

variable geometry — characteristic of aircraft with **variable sweep wings, swing wings**.

variable pitch propellers — the blades' angles of attack can be changed from coarse to fine. See **fixed-pitch propellers**.

variable sweep aircraft — an airplane equipped with a wing that may be moved from 90-degree angle to the body to a swept-back configuration, in flight as in the F-111. **Swing-wing**.

Vater und Sohn — see **Father and son**.

vectors — compass headings to reach a certain point, usually a **VOR** transmitter or an **intersection**.

vectors for sequence — term used by **ATC** to indicate that aircraft will be **vectored** until established in its proper **sequence** for landing.

Vee formation — formation of aircraft, usually three, shaped like the letter 'V". Also called the **vic**. Other formations include the "line-astern" and "line-abreast," but according to Monty Berger, during WWII such formations were "found to be easy targets," and "practiced squadrons" adopted the **finger four**, which see.

vegetables — sea mines, sown during WWII bomber **gardening** operations. —Doug Sample CD, Canadian Branch, Yorkshire Air Museum. In a logbook/diary kept by Capt. Sydney Wallis, then an RCAF flight officer, he refers to them as "carrots" and "onions" ("looks like we two crews are going to finish up planting vegetables tonight") and confesses to "getting sorta tired of the" gardening. But "our luck has been holding out well. Radar planting again—the other side of Denmark." (Quotation courtesy of Susan Rajsic, "No Price Too High" film project, Orillia, Ont., and before her, Capt. Wallis' widow).

venturi tube — a horn-like predecessor of the **pitot tube**.

vertical egg — "air manoeuvre in which two fighter planes chase one another in ever-widening loops." —Murray, "..Navy Fighter Pilots."

vertical reverse — "You fly straight and level and build up your speed, and then just do a tight turn, then pull back on the stick, all the way, and you're goin' backwards." —Williams, *The Plan*. **Tailstand**.

VFR — Visual Flight Rules. See **IFR**.

vic — **Vee formation**, shaped like the letter vee, from the old phonic alphabet, Victor.

victory roll — manoeuver performed over the base by returning fighter which had scored a "kill." See **clearance roll**.

Village Inn — familiar slang name for the airborne gunlaying turret, according to Jack Marriott, RAF.

VLCT — In 1994, design studies report on Very Large Commercial Transport, collaborative effort between Airbus Industrie of Toulouse and Boeing of Seattle, also glossed the meaning of the acronym as "Very Long Time Coming - VLTC", quoting Airbus president Jean Pierson. —*Interavia*, June 1994. The ICAO 50th anniversary book notes the use also of UHCA for "Ultra High Capacity Aircraft," in the realm of projected size, and in the realm of speed, both "HSCT" and "HST" for "High Speed Commercial Transport" and "Hypersonic Transport," not to mention "Orient Express," "an aircraft that follows a ballistic-like trajectory, becoming for a time a spacecraft before returning to earth." The sky appears not to be a limit to these concepts; perhaps it is left to sardonic slang to fill that role.

Voisin — early French flying and aircraft building brothers famous for their **boxkite**, **pusher** designs.

volplane — an archaic verb from French meaning "to glide," and as an noun, "the glide itself." "Suddenly airmen appear. They come down silently from a great height in far reaching volplanes." — Drew, *Canada's Fighting Airmen*.

VOR — VHF omnirange antenna, a **navaid**.

VOR-TACAN — VHR omnirange antenna with tactical air navigation - military version.

VTOL — vertical takeoff and landing, like the **jump jet**.

vulture — "a troublesome passenger," in commercial aviation. See also **hawk**. — Dickson, *Slang!*

WAAF — Women's Army Air Force. WWII. Early **aviatrix** included Amelia Earhart and Jacqueline Cochrane. See note under **RAF**.

WAG or wag — wireless air gunner — wireless operator. —Doug Sample CD, Canadian Branch, Yorkshire Air Museum. "We worked the radios for the navigation flights 24 hours a day. We were eight girls to a shift. WAGs in the a/c sent us their position every half hour, except at World Series time. Then they switched over to the broadcast band and we never heard from them the whole time till they got back." —RCAF ground staff, WWII. —Williams, *The Plan*.

wake turbulence — turbulence caused by the passage of an airplane through the air can be dangerous to following aircraft. Also called **engine wash, prop wash**.

wakey-wakey pill — tablet issued to aircrew to prevent drowsiness. WWII. —Doug Sample.

walkaround — same as **preflight**.

Walkers' Club — U. S. aircrew of the Asian Theater in WWII who had to return from missions behind Japanese lines on foot. **Evader**. See also **Caterpillar Club**.

walking into the treetops — early emergency landing strategy. On an early Fleet 50 Freighter piloted by Sheldon Luck, August, 1939, for Yukon Southern Air Transport (an early Grant McConachie venture)—"if he could just walk the Fleet in, slowly, tail heavy and let it settle among the tree tops [young aspen and pine...enough cover to cushion the impact...] letting them break its fall....this one didn't work out right. It killed the dog and ruined the aircraft." —Beaudoin, *Walking on Air*.

wand — term for the long tapered flashlight extension used to signal aircraft on the ground or a flight deck at night. Often they were different colors, indicating different members of the ground crew. See **Red Shirt, Blue Shirt, Brown Shirt, Yellow Shirt**.

wang — among hang glider flyers, a **wingover** manoeuvre making a sharp 90-degree turn.

warbird — any military aircraft, especially WWI or WWII aircraft.

warpaint — camouflage or other paint scheme used by the military in wartime on **warbirds**. See **nose art**.

washed out — failed, as in a flight training course. See **"cease flying."** Also can mean wrecked, as of an aircraft. **Scrubbed, grounded**.

washing machine — WWII use for a "flight commander's airplane for aviation cadet qualification flights;

a plane in which an unsuccessful cadet is 'washed out'." —*AAF: The Official Guide to the Army Air Forces.*

washing machine Charlie — "name given by my Marines on Guadalcanal to the twin-engine Japanese Betty bomber with intentionally unsynchronized engines that would circle over Henderson Field nightly, dropping random 250-lb. bombs, single-engine fighter-harassers were called 'Louie the Louse.'" —Col. R. Bruce Porter, *Ace! A Marine Night-Fighter Pilot in World War II.*

WASP — Women's Air Service Pilots, in WWII, American female flyers mostly used for ferry duty, including four-engine transatlantic bombers.

watch tower — Canadian and British WWII word for "control tower." —from Doug Sample CD, Yorkshire Air Museum, Canadian Branch.

Watch your 6 — stencilled taunt, in Cyrillic (Russian alphabet) characters, on the wing of a CF-18 on view at the St. Hubert, Québec air show in the early 1990s. It means "check out what's at six o'clock on your compass face, i.e. right behind you—we're on your ass!" See also **check six**.

water bomber — an aircraft which is used to fight fires. Early versions were the Consolidated PBY Catalina amphibian, helicopters carrying the **monsoon bucket,** the Martin Mars flying boat, and the Grumman Avenger (a landplane). More recently, the Canadair CL-215 and the turboprop CL-415 "scoop and drop" aircraft have so dominated the market that in France their nicknames provide the generic word-list for such aircraft: the 1982 *Petite Larousse* cites "le **Canadair**," but it is also known in Europe as "le **pelican**," and "avion citerne" (cistern airplane); and in California and Australia as the **Super Scooper.**

water wagon — an aircraft powered by a water injection turbojet engine, such as the Boeing B-52G.

wave-off — naval expression for the signal from the **batsman** to abort a carrier landing, usually due to a **foul deck** or a **missed approach**.

waving base — British expression for the area at an airport where those who accompany passengers congregate to wave them goodbye.

waxing tail — in dogfighting, getting and staying on your opponent's tail. —Chuck Yeager's autobiography.

weapon pack — (*waffentropfen*) weapon panier protruding from the bomb bay of a Junkers Ju-88, accommodating two 20 mm cannons. WWII.

weathercock — to turn into the wind, said of bush planes on floats, in lakes. **Weathervane**.

weather radar — ground mounted or onboard, used to "see" weather. Tuned to identify clouds, rain masses, and other weather phenomena. See **stormscope**.

weathervane — see **weathercock**.

wedding — "pairing up of one fighter plane with one **bogey**." —Murray, "Language of Naval Fighter Pilots."

weeds — "Indicates that aircraft are operating close to the surface." —Canadian Forces Aircom Regulation Glossary.

weekend birdman — flies only on weekends. Alternately; weekend pilot, weekend aviator, weekend sky jockey, etc. **Sunday driver**.

weekend warrior — military reservist or Air National Guard who typically flies only on weekends.

weight and balance — "seldom used, fortune telling by numerology." — Baxter, *More Bax Seats*.

weight on wheels, weight off wheels — in the first instance, describes an aircraft on the ground, and in the second, aloft.

wet compass — a compass that floats in liquid. See **whiskey compass**.

wet lease — arrangement to lease the aircraft with its crew and technician. —ICAO 50th anniversary book

wet or dry — see **fercht oder getrocknet**.

wet pee school — affectionate RCAF WWII name for WETP, War Emergency Training Programme, a school for aircrew to make up academic deficiencies. —Williams, *The Plan*.

wet point — underwing point where jettisonable fuel tanks (**drop tanks**) can be installed.

wet wing — a way of saying that the wing is also a fuel tank. You can have a tank within the wing, but a "wet wing" is a wing which is sealed so that it is itself a tank.

wet-winging — "a procedure was developed out of Viet Nam combat which was aptly called 'wet-winging.' A heavily damaged A-4 had made its way to the water, fuel streaming from its ruptured tanks. Finding the A-3 tanker, the Skyhawk pilot plugged in, and the two aircraft flew back to the carrier, the tanker feeding JP-4 [a type of fuel] continuously to the struggling A-4 as it approached the deck. Finally, just before touchdown, the A-4 disconnected and landed safely. Wet-winging became a standard procedure for damaged planes returning to the carrier." —Mersky and Polmar, *The Naval Air War in Viet Nam.*

whack or **bonk** — among hang glider flyers, to drop the nose of the craft sharply when landing, causing a crash. See also **prang.**

whale — slang term for **jumbo** jet. In the moments just before the collision (March 27, 1977) on the runway at Tenerife, Canary Islands, between a Pan American 747 and a KLM 747, a copilot shouted the warning, "There's a whale on the runway." See **wide body.**

wheel pants — wheel fairings. Also known as **trousers, skirts, pants, tin drawers, spats.**

wheels in the well — gear up. "We want to be wheels in the well by 8 o'clock" — i.e. in the air.

wheels-up — though technically the wheels don't get retracted until the aircraft is truly airborne, a slang term for exact moment of liftoff, when the wheels leave the ground is "wheels-up at......"

whirlybird — helicopter, **chopper, copter, egg beater, sling wing.**

whiskey compass — a compass with indicator that floats in liquid, usually alcohol. Though an old term, still in use by British Airways. See **wet compass.**

whiteout — a disoriented condition where a loss of a perceived horizon results from falling snow, blowing snow, sometimes caused by a helicopter downwash, or in bright sun over a large expanse of snow. See also **greyout, brownout, blackout, redout.**

white-tail — new aircraft, not necessarily white in colour, but without design, logo or distinctive colour (**livery**), sitting in the "parking lot" of the manufacturer, waiting to be sold.

whooping cough flights — French doctors in 1938 discovered a cure for infant whooping cough. "Physicians cannot explain why, but a short flight at an altitude of about 7000 to 9000 feet in an unpressurized airplane has cured a number of children. Worried parents can now enroll their offspring at local flying clubs for special 'whooping cough flights.'" — Gunston, *Chronicle of Aviation.* See **mercy flights, aerotherapeutics.**

whop prop — the helicopter rotor, nicknamed for the percussive sound it produces in flight, as anyone who has seen the invasion scene in the movie Apocalypse Now remembers.

wiblick — whimsical pronunciation of WBLC (Waterborne Logistic Craft) or potential enemy river traffic, by U. S. Navy pilots in Viet Nam. —Mersky and Polmar, *The Naval Air War in Viet Nam*.

wide body — generic name applied to modern generation airliners with two aisles (and, according to American Airlines training materials, may have up to three cabins.) See also **jumbo, narrowbody, whale.**

widowmaker — said of airplanes prone to disaster. Commonly applied to the Vultee BT-13 **Vibrator**. "We had only two cadets wash out in **Basic**, but we lost ten in fatal accidents. It was probably these statistics that gave the Vultee Vibrator its reputation of being a widowmaker." — Bill Kennedy Jr., *Air Classics Magazine*, May 1980. Also said of the Lockheed F-104 Starfighter.

wilco — radio parlance for "will comply."

Wild Boar — German night fighter operations that hovered over the target to attack the bombers from above. WWII. See also **Tame Boar**. —Doug Sample.

Winchester — "no ordnance remaining." (Out of ammunition).

—Canadian Forces Aircom Regulation Glossary. Also American use, in Vietnam, for "out of bombs, rockets, or bullets." —Col Bob Stoffey, *Cleared Hot!*

winco — short for Wing Commander, RAF.

windmill, windmill effect — a freely swinging fixed-pitch prop on a shut-down engine. Or, the propeller. —W. H. *What's the Gen?*" See **feather.**

window — thin strips of metal foil or metallic coated paper strips dumped from aircraft to foul enemy radar. Also known as **chaff, chuff.**

wind shear — violent downdrafts caused by temperature inversions and other weather phenomena. "Wind changing direction while aircraft is climbing or descending through altitude." See also **rotor.**

wind sock — "a cloth cone which indicates the wind direction," the **pants leg**. — Rose, *A Thesaurus of Slang*. See **tetrahedron, wind T.**

wind T — see tetrahedron.

wind tunnel — according to Learjet engineers, now known as "computational fluid dynamics" site.

wings — emblem awarded upon successful completion of flight training, a badge of honour and symbolic of a tough unwritten code

- don't get caught wearing them if you haven't earned them. See **leg spreader.**

a **wing and a prayer** — what pilots may find themselves flying on at some point.

Winged Boot — cloth patch worn among WWII military fliers who had evaded capture and "walked back." They were known as the **Flying Boot Club.** See also **Caterpillar Club, Walkers Club, Gold Fish Club.** —Douglas Sample.

wing heavy — a condition of rigging in which one wing tends to sink.

winglet — a small vertical upturn on the end of the wing, added for greater lift, stability and control, as on Boeing Boeing 747-400, Lear 45, and Canadair's Regional Jet. May also descend below the wing.

wingman — "Pilot of a plane led by another pilot, the **element leader.**" — From John and Adele Algeo, "Among the New Words: Gulf War Glossary," *American Speech* (Winter 1991). Second aircraft in a two-aircraft fighter formation, the lead aircraft called the "leader," wingmen normally flew behind, slightly to one side, and slightly higher. Their job was to protect the leader. "Wingmen were told that even if the leader flew into the ground the wingman must follow." — Deighton, *Fighter.*

wing-over — a flight manoeuvre in which the aircraft is alternately made to climb and dive during a 180 degree turn. According to Williams, *Casey Jones Cyclopedia of Aviation Terms*: A manoeuvre in which the plane is put into a climbing turn until almost stalled, the nose being permitted to fall during the turn so that a dive is begun in the reverse direction. A modification of this is the **hammerhead stall.**

wing walker — not only **stunt** performers on old biplanes but now also crew who serve as troubleshooters by walking along observing the wings during **cat launch** off aircraft carriers, and during manoeuvers of new aircraft around and out of hangars, to make sure nothing collides with anything.

wing walking — stunts performed by a daring acrobat on an aircraft's wings during the **barnstorming** era.

wing-warping — very early modification of aircraft design to try to find more control and stability, specifically in banking and turning, succeeded by the introduction of **canards, ailerons, winglets, empennage.**

witches' water — cynical nickname for aviation gasoline during WWI when being shot down often meant falling to a fiery death because of it. Also called **hell brew, infernal liquid, orange death.**

wobble pump — priming pump used to force fuel into cylinders prior to starting. Also called "primer."

Works and Bricks — In WWII military aviation, section responsible for the maintenance of airfield buildings. —Doug Sample.

WOXOF — "recreation time stemming from a grounded mission. (Yonay 1983 ['Top Guns,' *California* magazine, Vol. 8, No. 5, May 1983, pp. 94-102, 144-47] explains 'In aviation circles the letters and zeros in WOXOF refer to various unfavorable weather conditions—X means "obscured" for example, and F means "fog"—and strung together the letters mean "You're grounded, why not make the best of it?" —Murray, "Language of Naval Fighter Pilots." A "recreation room adjacent to the officers club in which one can usually find a bar, pinball machines, etc.," is a WOXOF room.

WRAF — Womens' Royal Air Force or a member thereof (pre 1947).

wrench-benders — slang for aircraft mechanics.

Wright brothers — Wilbur and Orville, first American heavier than air powered flyers.

Wrong-way Corrigan — "Federal aviation officials here, who believed that Douglas Corrigan was returning to California, were not amused with the explanation that 'his compass must have been wrong' when he landed [July 18, 1938] in Dublin instead.

"Already dubbed 'Wrong-way Corrigan,' the impish aviator may face disciplinary action on both sides of the Atlantic. He has violated US regulations by making an unauthorized ocean crossing and Irish ones by failing to obtain permission to land on Irish soil.

"Corrigan told workers at Floyd Bennett Field, Brooklyn [NY], that he was headed west. He got into the plane [a dilapidated 9-yr-old Curtiss Robin], secured the door with a piece of wire (it had a broken latch), took off and headed east — to the astonishment of those on the ground. For his nerve, Corrigan has been made a life member of the Wisconsin Liar's Club." —Gunston, *Chronicle of Aviation*. Today the term may be applied to anyone who gets lost or is prone to navigational errors. See also **red on blue.**

WSO, wizo, wise-o — weapons systems officer/operator.

WX, WIX — weather. **WX sno** is code for snow, **WX clg** for "ceiling," and **WX frz** for freezing rain.

X-planes — experimental aircraft such as the Bell X1, etc.

yahoo! — among hang glider flyers, the pilot's version of the paratrooper's "Geronimo!" Exclamation upon launching (usually off a precipice into the air....").

Yank drome — in WWII, an airfield operated by U. S. forces, according to the logbook of RAF F/O Sydney Wallis, a Canadian.

yarded — "I decided it was not going to leap off, so I yarded it off." — Spring, *Daring Lady Flyers*. A gradual, step by step take-off.

yaw — to turn toward left or right, from the point of view of the rear (where the rudder is). One of several air terms from boat talk. An "adverse aileron yaw" is the tendency of an airplane's nose to veer or yaw in the direction of the downward-moving aileron because it causes more drag than the upward-moving aileron, corrected with opposite rudder.

Yellow Shirt — U.S. Navy term for aircraft carrier crewman in charge of parking aircraft on the flight deck. So named because of his yellow shirt. See **Blue Shirt, Red Shirt, Brown Shirt**.

yoke — Control wheel used for turns, climbs, and descents. See **control yoke, joy stick**. "I was a twenty-six-year-old with a thousand hours of fighter time, who had almost died of disappointment the first time I saw the U-2. I looked in the cockpit and saw that the damn thing had a yoke, or steering wheel. The last straw. Either you flew with a stick like a self-respecting fighter jock or you were a crappy bomber driver—a damn disgrace—who steered with a yoke, like a damned truck driver at the steering wheel of a big rig." —Marty Knutson, quoted in Rich, *Skunk Works*.

you fly, you die — said of aircraft entering a **no-fly zone**.

Yugo — among "stewardesses," a slang term for a "small commuter plane," according to John Davis, *Buzzwords*. They also call them "lighter," "drop fly," and "treetop."

yum-yum yellow — pilots' word for color of the **Mae West**, WWII life-jacket, rumored to be appetizing to sharks.

zang — French expression for "an airplane."

zero decision height — In extremely foggy conditions, an aircraft may be permitted to land in zero visibility, if the runway is equipped with Category IIIc Instrument Landing System, "proper ground-guidance system, normally taxi-guidance lights and ground-control radar, to bring the aircraft to the gate after a foggy landing. Also, to disperse passengers, the airport must have a rail system that connects it to the nearest city." —"The Middle Kingdom, " *Toronto Globe and Mail* (February 1, 1996).

zig — "shoot down an airplane." —Murray, "Language of Naval Fighter Pilots." "Zigged" means "shot down."

zip-lip — radio silence. —Coonts, *Flight of the Intruder*.

Zipper — "Acknowledge radio transmissions with two clicks of the mike button." —Canadian Forces Aircom Regulation Glossary.

zoom — "The word zoom when used in speaking of airplanes means climbing at a very sharp angle. Usually an angle which is so steep, that the plane will **stall** if the angle is not changed in a few seconds. Some planes can zoom longer than others. Some of the high speeded [sic] racing planes can zoom at an angle of 90 degrees with the earth, for more than a thousand feet, while other planes cannot zoom more than a few hundred feet." — Dugal, *Readings on Aviation*, (1929).

zoom bag — "Flight suit." — From John and Adele Algeo, "Among the New Words: Gulf War Glossary," *American Speech* (Winter 1991).

zoomie — "Air Force pilot." — From John and Adele Algeo, "Among the New Words: Gulf War Glossary," *American Speech* (Winter 1991). Also noted to mean "naval aviators."

Zulu time — Greenwich Mean Time, from the dividing line or "0 hour" in England, and named from the "earth's time zones having once been divided among the letters of the alphabet." —Barry Lopez. Jocularly, time used by African pilots.

'PLANE NAMES

We do not aim to provide a complete list of aircraft names here; you may consult Gordon Wansbrough-White's *Names With Wings* (Airlife, 1994) and John Horton's *The Grub Street Dictionary of International Aircraft Nicknames, Variants and Colloquial Terms* (Grub Street, 1994).

As officially named by the manufacturer, the imagery of names includes many animals (Otter, Beaver, Buffalo, Bison, Bobcat, Tiger, Camel); many references to the heavens (Constellation, Galaxy, Starfighter, North Star, Jet Star, Lear Star, LodeStar); to bugs (Geoffrey de Havilland was an entomologist, hence Tiger Moth, etc.); words that associate with royalty (Beechcraft King Air, Queen Air, Baron, Duchess); many, many birds: Goose, Mallard, Widgeon, Falcon; and a range of hallucinatory phenomena such as the Dassault Mercure, Mirage, and Mystere, and the Russian Kamov Werewolf helicopter.

A number of aircraft were named for places, not all of them significant as locations of air activity: Lancaster, Blenheim, Halifax, Hampden, Albermarle, Wellington, Warwick, etc. Vickers tended to name its craft with words beginning with V or W. Lockheed is fond of star names.

We have included only such nicknames as refer to visible features of the aircraft or exhibit linguistic peculiarity and colour.

Aardvark — nickname of the General Dynamics F111 jet fighter-bomber, so named because of the distinctive shape of its nose.

Airbus — a family of advanced-technology airliners built in Europe by the French, British, German, Spanish consortium. A jocular nickname is **Scare Bus**.

2. a generic term applied, pre-Airbus Industrie, to any large transport. For example, huge fuselage, B-344 Stratocruisers were redesigned B-29 and B-50 aircraft with Pratt and Whitney piston/prop engines with turbo supercharging, giving them a "jet effect." Later variants known as **Guppy, Pregnant Guppy, Superguppy**.

Air Force One — also called "the Flying White House," this is the U. S. Presidential aircraft. Originally a Boeing VC-137C, more recently a Boeing 747, the VC-137C was equivalent to the commercial 707-320B. These aircraft are completely equipped for the conduct of presidential business, including, e.g., secure phone lines.

Albatri — name given by Allied forces to German Albatros fighters during WWI. Because there were so many variations, all Albatroses were collectively called Albatri.

Aluminum Overcast — Convair B-36. Ten engines, six props, four jets. First of the nuclear bombers, biggest combat plane of its time. Post WWII, 1946-7. It was also known as America's "big stick." The B-36 had such a long range and endurance that crews joked that a calendar was included on its instrument panel. The term "aluminum overcast" is also used humorously to describe a crowded sky.

Annie — nickname for the Avro Anson twin-engine trainer of RAF & RCAF WWII. The name was usually affectionate, the Annie being regarded as a safe and stable aircraft, sometimes referred to as "an overgrown Cub." Because of its docile flying characteristics, instructors would say to their students, "If you kill yourself in an Anson, you're better off dead." See **Flying Greenhouse**.

Army Mule — semi-official designation of the Piasecki H-25 single-engine utility helicopter of 1948.

Auntie Ju — nickname for the Junkers Ju-52 three-engine transport of WWII. See **Iron Annie, Tante Ju**.

Aurora — supposed successor to the Lockheed SR-71 **Blackbird** spy plane reportedly under development by Lockheed, a notion continually denied by both Lockheed and the Pentagon. However, an earlier Lockheed anti-submarine patrol aircraft carried the name.

Avro Arrow — twin engine delta-wing jet fighter of the 1950s, planned by the Canadian government for use in the Cold War. One built; shot down by the Diefenbaker government; some remains are in the Canadian Aviation Museum in Ottawa. Reportedly the most advanced aircraft of its time (the first **fly-by-wire**), the Arrow was a victim of, and is now synonymous with, "political casualty."

Baby Boeing — the Boeing 737, a twin-engine jet airliner, the smallest of the Boeing fleet. In Europe, known as the **City Jet**.

Bamboo bombers — USAAF gliders which delivered airborne troops in Burma and for D-Day 1944. "Had the glide angle of a crowbar." — Stevens, *There I was.* Also, nickname given the WWII Cessna T-50 Bobcat due to its largely wooden construction, and to the similar Cessna Crane post-WWII, according to Spring, *Daring Lady Flyers*.

Banana Boat — nickname for the Consolidated B-24 Liberator, because of its shape.

Bandit — nickname for the Embraer Bandeirante, a Brazilian twin turboprop 18-seat transport.

Bantam bomber — nickname of the Douglas A-4 **Skyhawk** light jet fighter-bomber, so named because of its diminutive size.

Barrel — nickname for the Saab J29 jet fighter because of its plump fuselage shape. In Swedish, **Tunnen**.

Bebe — Nieuport, diminutive but very effective French pursuit sesquiplane of WWI.

Big Bertha — "a thirty passenger Fokker airplane." — Rose, *A Thesaurus of Slang*.

Big European — the Airbus A-340, four-engine jumbo long-haul jet, consumes only 6 tons of fuel per hour, seats 262-335 pax, range 11,000 to 14,400 km. Costs $100 million. —Karl Morgenstern, *ILA News* (Berlin Air Show) June 16 1992.

Big Ugly — B-52 bomber — From John and Adele Algeo, "Among the New Words: Gulf War Glossary," *American Speech* (Winter 1991). See also **Buff**.

Bird-dog — Cessna O-1 observation/forward fire control and observation/liaison aircraft, c. 1949.

Blackbird — Lockheed SR-71 high altitude spy plane. Only aircraft to our knowledge given its letter-number designation by a presidential error: Lyndon B. Johnson accidentally reversed the letters in the press conference which finally acknowledged its existence: rather than embarrass the Man, they changed the RS-71 to the SR.

Black Death — Ilyushin IL2 **Stormovik**, as referred to by German ground troops.

Black Widow — Northrop P-61 WWII twin-engine radar-equipped night fighter, always painted black to reduce its visible signature.

Blenheim, Beaufighter, Bolingbroke, Beaufort — Bristol two-engine aircraft, used variously as fighters, reconnaissance craft, torpedo anti-submarine warfare, and bombers, and known also as "Blens," "Beaus," and "Bols."

Blitz — Heinkel HE-117 mail plane of the early 1930s. Blitz means "lightning" in German.

Bloody Paralyser — nickname for the Handley Page 0/100 and 0/400 twin-engine biplane bomber of WWI. The appellation derives from an order by the British Admiralty to Sir Frederick Handley Page to build 'a bloody paralyser of an aeroplane.' Powered by the first V-12 aero engine, built by Rolls-Royce adapted from the engine of the Silver Ghost.

"True, it was like a lorry in the air. When you decided to turn left, you pushed over the controls, went and had a cup of tea, and came back to find the turn just starting; but it had a nice comfortable feeling, leisurely and relaxed. If you had to drop bombs about the place, well, this was a gentlemanly way of doing it." —British author Cecil Lewis, quoted in Bowyer, *Handley Page Bombers of the First World War*.

Blue Canoe — Cessna model 310 U-3 twin-engine light utility transport first flown in 1954. So named because of its distinctive blue colour scheme matching that on USAF VIP transports. Also designated L-27. See **Skyknight**.

Bolly — a Bristol Bolingbroke aircraft (WWII **Blenheim** Mk IV).

Bolt, 'bolt — nickname for the Republic P-47 Thunderbolt. See **Jug**.

Boo — nickname of the de Havilland Canada C-7 **Caribou** twin engine STOL tactical cargo transport. First flown 1958.

Boomerang — Commonwealth CA-12, Australian-built WWII fighter.

Boston — Douglas twin-engine Douglas A20 bomber (Brit. term). Americans called them **Havoc**.

Brisfit — (RAF) popular name given the Bristol F-2 Fighter of WWI.

Britain First — see **Rothermere Bomber.**

BUFF — "Big Ugly Fat Fellow/Fucker, the B-52 bomber" — From John and Adele Algeo, "Among the New Words: Gulf War Glossary," *American Speech* (Winter 1991). See also **Big Ugly.**

Butcher Boy, Butcher Bird — nickname of the Focke-Wulf Fw-109 single-engine German fighter, WWII.

Buzz Saw — early nickname of the 747, according to John Wheeler, Boeing public relations, because of the sound made by the engines. See also **wide-body, jumbo.**

Camel — WWI Sopwith pursuit aircraft, so named because of its "hump-backed" silhouette, caused by machine-gun housings mounted on top of the cowling.

le **Canadair** — generic name in France for any water bomber, because of the success of Canadair's CL-215 amphibian, a "scoop and drop" firefighting aircraft. Listed with this meaning in the 1982 *Petit Larousse illustré* French dictionary as a "nom déposé." Also known as an **"avion citerne"** ("cistern airplane"),"**le pelican**," and latterly in California and Australia as the **Super Scooper.**

Some idea of the way this aircraft captured the popular imagination may be gleaned from a story of dubious authenticity which featured it: "there had been a forest fire near Grasse, and the Canadair planes had been called out. These operated like pelicans, flying out to sea and scooping up a cargo of water to drop on the flames inland. According to Faustin, one of the planes had scooped up a swimmer and dropped him into the fire, where he had been *carbonise.*" —Peter Mayle, *A Year in Provence.*

Canberra — Martin B-57 jet bomber, originally English Electric, the British Aerospace Corporation (BAC) Canberra. Exceptional in that this aircraft kept its original British name for the licence-built American versions.

Canso — Canadian version of the Consolidated PBY Catalina (as it was known in the US).

Caravan — Cessna 208 high-wing, single turboprop, cargo/ utility transport. Bulky and oversized, it has been called a "207 on steroids" and a "172 with thyroid problems."

Caravelle — the Sud Aviation (later Aerospatiale) four-engine jetliner. Known also as "the French Comet," after its competition, from de Havilland.

Cat, Catalina — Convair PBY amphibian flying boat, as called by the USN during WWII.. See also **Canso, Dumbo.**

Century-series jet fighters — advanced technology fighters beginning with the F-100.

Champ — the venerable Aeronca Champion, high-wing single-engine light plane, that pioneered recreational flying, post WWII. Aerobatic sportsplane. When Aeronca went defunct, the design was was later revived under the names Champion Traveller, Champion Decathlon, American Champion Scout.

City Hopper — European term for a commuter/shuttle aircraft.

City Jet — the Boeing 737. A total of 2,950 had been sold as of June 1992, making it "the most popular passenger aircraft in the world." —*ILA News* (Berlin Air Show), June 16, 1992. **Baby Boeing.**

CINC Hawk — nickname of the 'Commander-in-Chief Black Hawk', a Sikorsky H-60 twin turbojet helicopter specially equipped to accomodate U.S. Army commanders utilizing them as flying command posts. First developed in 1988.

Class of '47 — nickname for the Boeing B-47 Stratojet bomber because it was first flown December 17, 1947 and because it "was in a class by itself."

Clipper — Pan American's name for its planes. For a note on the connotations Pan Am hoped to lassoo with this word, see **ship.**

Cobra — "Two-person (tandem) helicopter used in attacking ground forces" —From John and Adele Algeo, "Among the New Words: Gulf War Glossary," *American Speech* Winter 1991). Later versions known as the Supercobra and the Cobra Venom. Bell AH1 helicopter gunship, also called the **Huey** Cobra since it was developed from the UH1 Huey airframe. Known to troops in Vietnam as the **Snake.**

Some examples of **code names** —

Backfire — NATO code name for the Russian Tu-22M3 supersonic medium bomber.

Betty — Allied code name for the G 4-ML Japanese bomber, WWII.

Bosun — "code name given a Tupolev-designed, twin-jet light bomber, somewhat larger than the **Butcher,** and first seen in quantity at the 1951 Moscow Air Show." — Heflin, ed. *The United States Air Force Dictionary.*

Camp — NATO code name for the Antonov An-4 four-engine transport.

Colt — NATO code name for the Antonov An-2 single-engine biplane transport.

Cub — NATO code name for the Antonov An 12 four-engine transport.

Fagot — NATO code for MiG 15 fighters ("F" for fighter).

Foxhound — NATO code name for the Russian MiG 31, all-weather interceptor.

Frogfoot — "Soviet-supplied Iraqi SU-25 attack plane." —From John and Adele Algeo, "Among the New Words: Gulf War Glossary," *American Speech* (Winter 1991).

Kate — Japanese torpedo bomber, WWII.

Mavis — Japanese four-engine flying boat, WWII.

Oscar — Allied code name for the Japanese Nakajima Ki-43 fighter, similar to but inferior to the Zero. Official name was Hayabusa.

Rufe — Nakajima A6M2-N Japanese fighter.

Val — Japanese Aichi 99 dive-bomber of WWII.

Zero, Zeke — Allied code name for Mitsubushi A6M, WWII Japanese fighter. Official name: Rei Sentoki (Japanese pilot's nickname: Reisen.)

Comet — first passenger jetliner, British, with pressurized cabins, etc.—many firsts. The fleet was grounded after two of them disintegrated in the air in 1954, and the Boeing 707 stepped into the market niche thus created.

Compass Call, EC-130H — **Jammer** plane — From John and Adele Algeo, "Among the New Words: Gulf War Glossary," *American Speech* (Winter 1991). Named for its mission, which is to jam transmissions essential to location and navigation.

computer with an airframe — description of McDonnell/Douglas F-18 Hornet, in Canada the CF-18, first Canadian military aircraft to rely on computer control (**fly by wire**). —P. Hargrove, Canadair.

Concordski — nickname of the Russian version of the supersonic transport (Tu-144). Crashed at Paris Air Show.

Connie — nickname of the Lockheed Constellation four-engine airliner with three vertical stabilizers (tail-fins) and in which the whole shape of the fuselage is an airfoil.

Converter — T-37 Cessna, because "they convert fuel to noise." Also known as the **Tweety-Bird**. Any loud, inefficient airplane.

Corsair — Chance Vought F-4U single-engine **inverted gull wing**, carrier-based fighter, ground support aircraft developed in 1938. Called the "finest fighter of WWII." The name had been used earlier by Chance Vought for a biplane observation aircraft of the 1920s. Unusual because the name was recycled twice: now it is applied to a Chance Vought navy jet.

Cub — Piper J-2 and J-3 single-engine, fabric-covered high-wing two-place tandem sportsplane of the early 1940s. Improved versions are known as Super Cub. "Cub" is often used generically for any small private aircraft or airplanes with docile flying characteristics. Thus the Avro Anson is sometimes called "an overgrown Cub." That term has likewise been applied to the Douglas DC-3.

Dakota — British and Canadian name for the Douglas DC-3/C-47 twin engine transport. See **Gooney Bird, Skytrain, Firefly, Grand Old Lady, Old Bucket Seats.**

Delta Dagger — Convair delta-wing jet fighter F-102. First flown in 1953.

Destroyer — nickname for the Messerschmitt Bf-110 twin fighter-bomber. Both the Luftwaffe and the Allies used the nickname. Simultaneously, the official name "Destroyer" was applied by the Allies to the Douglas B-26 bomber, later by the USAF in Viet Nam to the Douglas B-66 jet bomber, and the appellation has been similarly used for a number of aircraft since early days.

Devil Sled — nickname of the Messerschmitt Me-163 **Komet, Comet.** Tailless, so named because it was launched from a jettisonable wheeled sled and landed on skids with no shock absorbers.

diesel neuf — jocular nickname for the DC-9, among bilingual Air Canada pilots.

Dollar-nineteen — nickname for the Fairchild C-119 **Flying Boxcar.**

donut on a rope — nickname of the **Aurora**, experimental reconnaissance aircraft, capable of Mach 10, and powered by a pulse detonator wave engine which would put out recurrent puffs of exhaust punctuated by a repeated boom, and thus made a trail that looked like the rope on which the "donuts" were suspended. —Don Hackett.

Doodlebug — British term for the V-1 rocket, the "German flying bomb" or "buzz bomb." — Heflin, ed. *The United States Air Force Dictionary.* Also a pilotless aircraft, post WWI, radio-controlled, built by Charles Kettering (1919-21).

Dragon Lady — nickname of the Lockheed U-2 high altitude spyplane.

Drut — USMC nickname for the Douglas EF-10B Skyknight jet used in Viet Nam. The humor comes from spelling the name backwards. The Skyknight originated in Korea as the USN F-3D, the nickname possibly originating there.

Dumbo — affectionate name among US Navy aviators during WWII for the Consolidated PBY Catalina, also known as the **Cat**.

Dyna Soar — Boeing X20 experimental high altitude aircraft.

Electra — originally the Lockheed 10A twin-engine airliner of the mid-30s. Lockheed later reused the name for their four-turboprop airliner, first flown in December 1957.

Emil, **Emmy** — common name used by Luftwaffe for the Messerschmidt Bf-109E fighter of WWII. Although specifically applied to the 'E' model, the appellation was generally carried over to all models, but see **Friedrich, Gustav**. The names of these three letters are taken from the German phonic code (like "Alpha, Baker, Charlie")...."Emil, Friedrich, Gustav.")

Empress of — Canadian Pacific Airlines' name formula for its early airliners, following its parent Canadian Pacific corporation practice of naming ships. For the connotations, see **ship**.

Enola Gay — a specific Boeing B-29, piloted by Col. Paul W. Tibbets, named after his mother, which dropped the world's first atomic bomb on Hiroshima, at 9:15 a.m. August 6, 1945.

Eversharp — nickname for the WWII German Dornier Do-17 bomber also known as the **Flying Pencil** because of its long slender fuselage.

Fairey Swordfish — see **Stringbag**, more often **Swordfish**.

Faith, Hope, and **Charity** — nicknames of the three aging Gloucester Gladiator biplanes that defended Malta heroically during World War II.

Fat Albert — wide-body jumbo jet (Air Force slang). — Dickson, *Slang!*

2. Nickname applied by Bill Lear to the Canadair Challenger. It was "Bill Lear's moniker for Halton's redesigned LearStar. He also called it 'Fat Albert with a Nose Job' at a Canadair luncheon. It was to become the Canadair Challenger

bizjet. Lear called his design 'Allegro'". —from Logie, *Winging It*. See also **Mickey Mouse Learjet**. Of course the term "Fat Albert" comes from a character on the Bill Cosby show.

Fat Cub — nickname of the Piper Vagabond. See **Wagabond**.

Fat Face — US Navy codename for the Consolidated PB4Y flare dropping aircraft used in Korea. The PB4Y was a naval version of the B-24 Liberator but sported a single vertical stabilizer.

Fee — nickname for the British FE-2 biplane of WWI.

Firefly — name given Douglas C-47s equipped to drop flares for night air-to-ground operations in Korea. See **Dakota**, **Skytrain**, **Grand Old Lady**, **Old Bucket Seats** and **Gooney Bird**.

Flagship — American Airlines name for its planes.

flaming coffin — nickname given the de Havilland DH 4 biplane of WWI. American-built examples were known as the **Liberty Plane**. See **Swallow**.

Flanker — Russian SSU, Mach 2 Russian transport plane which, it's proposed, should be developed in a joint venture with Gulfstream to become the world's second commercial supersonic aircraft, after the Concorde. Also, and earlier, the NATO designation of the SU-27 twin engine fighter. A recycled name, we believe.

Flyer — the original Wright brothers aircraft.

Flying Banana — nickname for the Piasecki HRP Rescuer helicopter, so named because of its distinctive banana-shaped fuselage.

Flying Barrel — Prior to WWII Curtiss favored the use of a landing gear that retracted into the sides of the fuselage, such as on the USN FF-1, F2F and F3F carrier-based torpedo bomber biplanes. This resulted in a deep, slab-sided appearance and they were universally dubbed Flying Barrels.

Flying Bathtub — nickname applied to at least three different aircraft: the WWI Caudron, the Aeronca C (also known as the "Airknocker"), and the M2-F3, wingless and engine-less research aircraft designed by NASA (early 1960s) to study potential spacecraft designs. The M2-F3 was a craft that was dropped from another aircraft, and control was maintained through high dive speeds to flare and landing.

Flying Bedstead — nickname of a Fairey-designed, Rolls-Royce-powered craft which did very early vertical take-off, powered by two R-R

Nene engines and first flown by R-R test pilot Ronnie Shepherd.

Flying Boxcar — popular name for the Fairchild C-119 twin engine cargo transport. Also known as the **Dollar-nineteen**. Earlier applied to the Junkers Ju-52 **Iron Annie**. Also applied to the earlier Fairchild C-82, which was officially known as "The Packet."

Flying Cigar — Convair B-36, for the long, slender shape. See **Aluminum Overcast**.

Flying Coffin — nickname for the Culver Cadet, single-engine low-wing cabin monoplane, never properly trimmable.

Flying Edsel — Convair 880/990 jet airliner project, which lost General Dynamics $425 million, the largest in American history to that time (mid-'sixties), more than Ford!

Flying Eye — Focke-Wulf Fw-189 twin-engine, twin-boom dive bomber/reconnaissance aircraft of WWII, so named because of its totally glassed-in nose section.

Flying Forest — what Charles Lindbergh, then technical adviser to Pan American Airways, called the Sikorsky S-40 flying boat, because of its incredible array of struts and bracing wires. The designer and builder, Igor Sikorsky, called it a "thing of joy and beauty."

Flying Fortress, Fortress, Fort — Boeing Model 299 B-17 long range heavy four-engine bomber, WWII. One German fighter pilot said, "Attacking a B17 was like making love to a porcupine."

Flying Fuel Tank — Boeing B-29 **Superfortress**, after setting a distance record in 1945.

Flying Greenhouse — nickname of the Heinkel He-111 twin engine bomber of WWII, because of its long glassed-in canopy.

2. nickname for the Avro Anson twin-engine bomber of RAF/RCAF during WWII because of the versions which had an enormous amount of glass in the fuselage, turrets, and cockpit. See **Annie**.

Flying Handcart — (Brit.) Lockheed P-38 Lightning, because of its shape, with a double-boom tail. WWII.

Flying Houseboat — Italian designer Gianni Caproni envisioned a 100-passenger flying boat capable of transatlantic travel in 1919. His huge Model Ca-60 featured a huge houseboat-like hull and a unique arrangement of three sets of triplane wings, each supported by dozens of struts and braced by miles of wire. According to one air historian, the result would, "not have looked out-of-place sailing up the English Channel with the Spanish Armada in 1588."

Flying Laboratory — Douglas XB-19 used to test radical new design approaches in four-engine long-range bombers in the late 1930s. Also, a generic name applied to any test or experimental aircraft.

Flying Newsroom — name given to a de Havilland Dragon Rapide purchased by the Toronto *Globe and Mail* in 1937, to be used primarily to fly reporters to the Canadian North. Unfortunately it burned while refuelling at Toronto after only one flight to Northern Ontario.

Flying Pancake — Chance Vought V-173 and XF5U-1 experimental flattened-flying-wing-like aircraft.

Flying Pencil — nickname for the German WWII Dornier Do-17 twin engine bomber, also known as the **Eversharp** because of its long, slender fuselage. Also applied to the Junkers 88 and Heinkel 111.

Flying Prostitute — the Martin B-26 Marauder, the **Mystical Marauder**, also dubbed the "Flying Prostitute," because "it had no visible means of support." — Stevens, *"There I was."*

Flying Razor — Fokker D.VIII parasol wing monoplane of WWI, so named because of extremely sharp fuselage edges and angles. Last of the WWI Fokkers.

Flying Shoe — nickname given the Blohm und Voss BV 138 three-engine German flying boat of WWII. The aircraft gained a reputation for being virtually indestructable and particularly suitable to long-range maritime patrol. One story goes that an Allied convoy was being shadowed by a Flying Shoe which circled just out of range of the anti-aircraft guns. A sailor signalled by Aldis lamp, 'Could you stop doing that, we are getting dizzy.' The flying boat obliged by reversing course - and circling counterclockwise.

Flying Suitcase — Handley Page Hampden, a twin-engine bomber of WWII, so called because of its shape. See **Hamp.**

Flying Tank — Sikorsky Black Hawk helicopter. Also applied to the Ilyushin IL-2 **Stormovik** of WWII.

Flying Wing — The Northrop YB experimental bomber, stealth aircraft, pusher engines.

Fokker — Anthony Fokker, Dutch aircraft designer and builder. Returned fighter pilot, describing aerial combat for a mixed audience after the war: "And I looked up, and there was one fokker to the left, and one fokker to the right....."

Woman host breaks in...."I must explain that the 'Fokker' was a type of German fighter plane!"

Fighter pilot: "No, no, these fokkers was flying Messerschmitts!"

Fork Tailed Devil — Japanese nickname for the Lockheed P-38 Lightning of WWII, so named because of its twin-boom tail arrangement; even in German, **Teufelschwanz, Gabel-schwanz Teufel.**

fork-tailed doctor killer — irreverent nickname for the Beech Bonanza. "They're tricky to fly, and doctors make very poor aircraft student pilots because they tend not to listen to anyone." —Beverly Howard (Austin, TX, owner of Windermere Gliderport.)

Fort — Flying Fortress.

Fred — USAF nickname of the Lockheed C-5 Galaxy cargo jet, acronym for "Fucking Ridiculous Economic Disaster." —from a tanker pilot.

Freeman's Folly — first nickname of the **Mossie**, de Havilland Mosquito, so dubbed because the British Air Ministry turned down the design and it proceeded as a de Havilland private venture, Sir Wilfred Freeman being the concept's chief proponent.

Friedrich — German nickname for the Messerschmidt Bf-109F fighter of WWII. See **Emil, Gustav.**

Frying Pan — nickname for the USN Grumman A-6 Intruder, during Viet Nam war. So named because of its distinctive tapering fuselage. Has been described as "singularly unattractive." Also known as the **Tadpole.**

Gabelschwanz Teufel — German for "fork-tailed devils," P38 fighters.

Ghost — Saudi name for the Lockheed F-117 **Night Hawk.**

Gimli glider — Air Canada Boeing 767 that ran out of fuel and had to do a **dead-stick** landing at a small airstrip in Gimli, Manitoba. The problem was in the conversion (government-mandated) of fuel measurement from the Imperial system to metric. Control was possible because of the **rat**, and the fact that one pilot had a lot of glider experience, and the other knew where the old airstrip was. As a drag race was taking place, one car at the ready for burnout was very surprised at what passed just overhead.

Gladbag — RAF nickname for the Gloster Gladiator, last of the RAF biplane fighters of the late 1930s.

Global Express — this projected intercontinental bizjet by Bombardier/Canadair offers a smorgasbord of features which give some shape to the slang of the future: flat-panel EFIS (Electronic Flight Instrumentation System) displays, **HUD**, electronic library, paper-free cockpit, **fly by wire.**

Goblin — XF-85, a "**parasite**" jet fighter attached to a bomber mothership - experimental.

Gooney Bird — nickname for the Douglas DC-3/C-47. Also known as **Grand Old Lady, Dakota, Firefly, Old Bucket Seats** and **Skytrain**. Reappeared in Viet Nam, according to Bruce Callander, as "Puff the Magic Dragon."

Grand Old Lady — nickname for the Douglas DC-3/C-47 twin engine transport. Also known as **Gooney Bird, Dakota, Old Bucket Seats, Skytrain** and **Firefly**.

Gray Goose — the Grumman Goose.

die **gross arschen vogel** — "the big ass bird", German nickname for the Martin B26 Marauder, because it has a large tail.

Gun Bus — nickname for the Vickers FB-5 biplane pusher fighter of WWI.

Guppy — nickname for the On Mark, Conroy, and Canadair PG conversions from the Boeing B-77 Stratocruiser airliner of the 1950s. The Guppy has a huge fuselage which hinges open to allow the loading of oversize cargo. Also known as **Pregnant Guppy, airbus.**

Gustav — German nickname for the Messerschmidt Bg 109G.

Gypsy Rose Lee Warhawk — Curtiss P-4lL stripped of armour, armament, etc., for higher speed. Named for the famous ecdysiast (stripper).

Halibag — nickname for the Handley Page Halifax bomber of WWII.

Hamp — the Handley Page Hampden. See **Flying Suitcase.**

Harry Tate — nickname for the Royal Aircraft Factory RE-8 fighter of WWI, so-named because the airplane had a tendency to do some tremendously funny trick at the most unexpected time, just like Harry Tate, a popular London stage comedian of the time.

Harvard — British name for the North American T-6 trainer, called by the Americans **Texan,** and the Canadians, **Yale.**

Havoc — see **Boston.**

Hawk — the Curtiss single-engine biplane P36 fighter (pre WWII).

"The ancestral name of Hawk aircraft comes from Kitty Hawk in North Carolina, known throughout the world as the birthplace of powered flight." — Vader, *Pacific Hawk*. Kitty Hawk is where the Wright Brothers did their famous first flight in 1903.

Herk, Herky Bird— Lockheed C-130 Hercules four-turboprop tactical transport, USAF, c. 1954. According to *Air & Space Magazine*, March, 1996, the following are also Hercules nicknames: Snoopy, Fat Albert, Pinocchio Nose, Ski-bird, Vomit Comet, Eagle Owl, and Rhinoceros.

Hollering Huey — name for the Bell UH-1 **Huey** helicopter equipped with loudspeakers which could be heard for a mile and used by the USMC in Viet Nam.

Hoover — nickname for the U.S. Navy Lockheed S-3A Viking anti-submarine aircraft, so called "because it sounds like a vacuum cleaner when its throttles were jockeyed, especially on landing." —Wilson.

Huey — Bell UH-1, HU-16 jet-powered helicopter. "Huey became synonymous with mobility, transport, assault, dust off, and medevac in the U.S. Army and Marine Corps lexicons during the Viet Nam war. Thousands of the Bell-produced machines served in the conflict. The name Iroquois was officially assigned, but the helicopter was always called Huey, the name derived from the pre-1962 designation letters HU." —Mersky and Polmar, *Naval Air War in Viet Nam*. Gunship version, the AH1 is called **Huey Cobra, Cobra**, or **Snake**.

Hunchback — Ilyushin IL2 **Stormovik**, because of its shape.

Hurri — Hawker Hurricane, WWII single-engine fighter.

Hurri-bomber — Hawker Hurricane used in fighter-bomber role.

Hurryback — WWII slang for a Hurricane fighter. —W. H. *What's the Gen?*"

Illusion — Ilyushin 62, Russian aircraft. Copy of the Vickers VC-10, known as the **Classic** (NATO reporting code).

Iron Annie — nickname for the Junkers Ju-52 three engine transport, probably so-named because of its characteristic corrugated metal construction, and its virtual indestructability. See **Auntie Ju**.

Jayhawk — the nickname of several different aircraft, no doubt originally from Kansas (the Jayhawker State) because of the excellent flying conditions (flat land, clear skies) a major aircraft construction locale (especially Wichita—Boeing, Beech, Cessna, Lear). A particular Sikorsky Blackhawk helicopter involved in a daring rescue off the Atlantic coast was so named, its crew honored by the Aviation/Space Writers Association at the 1992 newsconference in Montréal. Also the nickname of the Beech business jet developed after a Japanese firm had failed to do so, during the mid 1970s, at a time when Japanese technological and business triumphs over U. S. were legion (the jet was successfully developed in Wichita, Kansas).

Jenny — WWI Curtiss biplane trainer, from Navy designation JN-3, 4; later a favorite **barnstormer** aircraft, from war surplus.

Jew canoe — a slang term for the Westwind, an Israeli-built aircraft. Also **Kosher Klipper.**

Johnson City Windmill — a Sikorsky S-51 helicopter used by Lyndon Baines Johnson in his campaign for the U. S. Senate in 1947, a dramatic contrast to the rail-borne political campaigns of the past. He used a loudspeaker and made visits to many Texas locales, astonishing many local voters who had never seen anything like it.

Jolly Green Giant — nickname for the Sikorsky HH-3 helicopter. Sometimes simply known as Jolly.

John Wayne Airplane — nickname of the Airbus, because "it goes into mountains, chops down trees, and kills Indians." —Jeff Heinrich, "Popular 'John Wayne' Airbus in no danger of flying into sunset," Montréal *Gazette*, June 17, 1994. The allusion to Indians refers to the crash in February 1990 in Bangalore in southern India, killing 92, the first crash of an Airbus on a commercial flight. According to Heinrich, it is also known as the **Fifi,** "the French Airplane with a Mind of Its Own," and **Christine,** for the runaway car in the Stephen King novel and movie.

J-stars —"converted 707 equipped to function as a surveillance aircraft with radar to spot armored vehicles and other ground forces." —John and Adele Algeo, "Gulf War Glossary," in "Among the New Words," *American Speech.*

Jug — nickname for the Republic P-47 fighter of WWII, so called because of its squat fuselage lines. Also, a piston cylinder, as in car talk.

Jump Jet — the Harrier VTOL jet fighter.

Kabam Saddam — nickname of one Canadian Forces chopper used in the Gulf War.

Kingfisher — Chance Vought shipboard catapult-launched aircraft of the 1930s.

Kitty-bomber — Curtiss P-40 fighter when used in fighter-bomber role. Punsters called it the B-40.

Kosher Klipper — Westwind 1124 or Astra Jet. Built by Israeli Aircraft industry. Also known as the **Jew canoe**.

Lanc — nickname for the Avro Lancaster bomber, WWII.

Learjet — specifically any of a line of executive jets designed and built by Bill Lear and later, Learjet division of Bombardier Aerospace, but also a generic term for business jets.

Lib — Consolidated B-24 Liberator four-engine heavy bomber of WWII.

Liberty Plane — nickname given the de Havilland DH-4 day bomber of the 1930s. The American-built versions were so-named because they were powered by a 425 hp Liberty V-12 engine. This aircraft earned the unenviable name **flaming coffin**.

Lightning — Lockheed Model 22 P-38 twin-engine, twin-boom fighter of WWII first flown in 1939. The name was bestowed by the British, who placed the first orders. The aircraft delivered to England were non-turbocharged and exhibited less than expected performance, thus giving rise to the appellation "Castrated Lightning." The two-seat trainer and night-fighter versions were known as **Piggy-Back**, because of their tandem seating. German pilots called them **Fork-Tailed Devils**. Japanese pilots referred to them as **Two Fighters With One Pilot**. Also, the English Electric jet fighter of the 1950s, and also the Heinkel He-117 **Blitz**.

Limping Annie — nickname for the Avro Anson aircraft. WWII. —Doug Sample. Also known simply as **Annie**.

Linc — nickname for the Avro Lincoln bomber, last of the piston-engine bombers to fly with the RAF during late WWII.

Little Bird — nickname for the Hughes 500M helicopter, an armed version of the H-6 Cayuse, used by members of the "special operations community."

Lizzie — (RAF) Westland Lysander liaison aircraft of WWII. —Doug Sample CD, Canadian Branch, Yorkshire Air Museum.

Long Beach Cable Car — the McDonnell-Douglas DC-9, because so many systems had cables and pulley backups, e. g. the cabin pressure relief valve is electronic, but it has cables also. Built in Long Beach, CA. —John Race, Canadair Challenger pilot.

Mainliner — United Airlines' name for its planes. The airline is famous for a long-running ad slogan: "Fly United."

Maytag Messerschmitts — P-22s, according to Stevens, *"There I Was...."*. Probably applied because the P-22 was not a successful aircraft. See also **Messerschmitt Maytag** in main section.

Mickey Mouse Learjet — one pilot's derisory name for Canadair-evolved LearStar, which became the Challenger bizjet. —from Logie's *Winging It*. See also **Fat Albert**.

MiG — acronym formed by the last names of two major Soviet aircraft designers, Mikoyan and Gurevich. Usually officially spelled MiG, but popularly, Mig.

Mig Master — nickname for the Grumman F-8 Crusader jet fighter used by the USN in Viet Nam, so dubbed by Public Affairs officials after considerable success against MiG 17s and 21s.

Millipede — Arado AR232 twin-engine German transport of WWII, so named because of its 11 idler wheels located beneath the belly to help support heavy loads.

missile identifying letters — "A" was the missile identifying letter for Russian air-to-air missiles; air-to-ground were called "K", where Russian missiles' names were not known. Similarly WWII Japanese planes, when their names were unknown to the enemy, were called by code words like **Zeke** (for Zeros), **Betty** bomber, etc.

missile with a man in it — descriptive term applied to the Lockheed F-104 Starfighter, which first flew in 1956. Unfortunately, this aircraft has also earned the appellation **widowmaker**.

Mitchell — bomber named for General Billy Mitchell, USAAF, a North American B-25 twin-engine attack bomber, made famous by the movie "Thirty Seconds over Tokyo," depicting the daring carrier-launched raid made by the Americans just after Pearl Harbour.

Mossie — de Havilland Mosquito, WWII fighter-bomber, plywood. The name reflects Geoffrey de Havilland's interest in entomology. See **Moth**. Also known as the **Wooden Wonder, Freeman's Folly**.

Moth — Name given to a long series of biplanes, mostly 1930s, by designer Geoffrey de Havilland, who was an entomologist (student of bugs): Moth, Moth Minor, Tiger Moth, Puss Moth, Dragon Moth, Gypsy Moth, et al.

Mystical Marauder — "Martin B-26 Marauder, the **Flying Prostitute..**" — Stevens, *"There I was."*

Mythhawks — see **Tomorrowhawks.**

Nancy — "a seaplane of the NC type." — Weseen, *A Dictionary of American Slang.*

nearjet — insulting name for the Cessna Citation at the beginning. "Risk of bird-strikes from the rear." — Logie, *Winging It.* Also known as the **Slowtation.**

Night Hawk — Lockheed F-117 stealth fighter employed on "covert surgical strikes." The entire aircraft is designed to produce a minimal visual, radar, and infrared signature, making detection almost impossible. In its early stages it appeared prone to crashes, giving rise to the name **Wobbly Goblin.** Saudi forces called it **Ghost.** Its prototype went by the code name **Senior Trend,** and an earlier test model was called **Half Blue.**

Noisy Northstars — nickname given the Canadair DC-4M conversion by Trans Canada Airlines president Gordon McGregor, according to Logie, *Winging It.* Excessive noise was produced by Merlin engines.

Old Bucket Seats — nickname given the Douglas DC-3/C-47 twin engine transport. Also known as **Gooney Bird, Grand Old Lady, Dakota, Skytrain** and **Firefly.**

Old Man's Cadillac — the Boeing 747, so dubbed by veteran Pan Am captain Ned Wilson in *For Pilot's Eyes Only.*

Ol' Number One — name given to a Helio Courier, N 242-B, the first one ever built, presently operated by **JAARS.**

Oxbox — nickname for the Airspeed Oxford twin-engine transport/trainer of RAF WWII.

Pandora — ten passenger airliner version of the Huanquero. Only someone who does not know the classical meaning of this word would name an aircraft after Pandora's Box!

Peashooter — nickname given the Boeing P-26 fighter circa 1933. Probably so-named because of the long peashooter-like gunsight or

pipper mounted atop the cowling forward of the cockpit.

Pelican — French nickname for the Canadair CL-215 and 415 firefighting water bomber, which scoops water which it drops on fires. Also known as **le Canadair, Super Scooper.**

Photo Joe — name given to Japanese **recce** planes (WWII).

pigboat — mechanics' term for the Lockheed L-1011.

Piggy-back — two-seat version of the Lockheed P-38 **Lightning.**

Pointer — "Nine-pound, remote-controlled model airplane with a TV camera in its nose, used for reconnaissance in the field." — John and Adele Algeo, "Among the New Words: Gulf War Glossary," *American Speech* (Winter 1991).

Pregnant Guppy — see **Airbus, Guppy.**

Puff — short for "Puff the Magic Dragon," originally AC-47 gunship (**Gooneybird**), now AC-130. Bristled with machine guns. —from Paul Turk. Also known as **Spooky.**

push-me-pull-you — nickname given the Cessna O-2 Skymaster because of its pusher-puller engine arrangement.

Quiet Trader — BAE-146, a small four-engine low-noise cargo/transport, capable of carrying 45-50 people.

Rhinoceros — F-4 Phantom. Pilot is known as a **rhinodriver.** Nickname also applied to the Lockheed C130 Hercules.

Rosinenbomber — "raisin-bomber," or "currant-bomber", Berlin slang for the aircraft that brought food during the Berlin blockade of 1947.

Rothermere Bomber — "In the years between the end of the Great War and the beginning of the 1930s very little progress had been made in the design and performance of RAF aircraft, and it was becoming increasingly evident that other countries — and potential enemies at that — were making greater strides than was Britain....until Lord Rothermere decided to take a hand.....This enthusiast and patriot commissioned the Bristol Aeroplane Company to produce a modern high-speed aircraft with a view to stimulating interest generally and so encourage the production and development of a number of up-to-date aircraft with which to equip the RAF.

The aeroplane produced by Bristol was a low-wing, twin engined cabin monoplane which would be deemed an 'Executive' type today, but in those days rejoiced in the names of the 'Rothermere Bomber' or **'Britain First'** — which latter name was more suitable as the aircraft was in no way a bomber." — Kent, *One of the Few.*

San Antonio Sewerpipe — Swearingen Metro or Merlin commuter aircraft, built in San Antonio TX, with the shape so described.

Sandy — nickname for the Douglas A-1 Skyraider, operated by the USAF in Viet Nam, known to the Navy as **Spad.**

Scare Bus — irreverent nickname for Airbus, the European consortium advanced-technology jumbo jetliner.

Schnell Bomber — name applied by Adolf Hitler to his fleet of "fast bombers" ("schnell" means "fast, quick" in German), twin-engine, medium range, medium-weight machines employed against Germany's neighbouring countries, notably the He-111, Ju-88 and Do-17.

Secret Bomber — Heinkel He-111 twin-engine German bomber of WWII, so named because it was developed in 1934 originally as an airliner for Lufthansa because Germany was restricted by the Versailles Treaty from building military aircraft. The name was applied in 1939 when its real mission was revealed.

Seven Seas — Douglas DC-7C long-range airliner of the late 1950s.

Shack — Avro Shackleton four-engine bomber, later AEW type.

Shorthorn — popular name for the Maurice Farman trainer, pre-WWI.

Short-Mayo Composite — Idea developed in Britain in 1937 by Major R. H. Mayo, of joining two aircraft, one atop the other. The larger, bottom aircraft would take off with the smaller one on top. When altitude and distance were attained, the two aircraft would separate, the smaller going on to its destination, the "mother ship" returning to base. The mother ship, a Shorts Empire flying boat christened Maia, was modified to carry a Shorts high-wing four-engine monoplane named Mercury. The first transatlantic crossing was made on July 21, 1938, from Foynes, Ireland, with Mercury carrying an enormous load of newspapers, mail, and fuel. Flying well out to sea, the aircraft separated cleanly; Maia returned to Foynes and Mercury flew safely to Montreal, covering the 2,930 miles in 20 hours 20 minutes. —Allward, *Il-*

lustrated History of Seaplanes and Flying Boats.

shufti-kite — RAF nickname for Ju-86P, high altitude **recon** aircraft.

Sky Crane — popular name for the Sikorsky CH-54 Tarhe twin-engine heavy-lift helicopter developed in 1958. Tarhe is a Wyandot Indian word meaning "crane."

Skytrain — nickname for the Douglas DC-3/C-47 twin engine transport, especially those used to drop paratroops during the invasion of Normandy. See also **Gooney Bird**, **Grand Old Lady**, **Dakota**, **Old Bucket Seats**, **Firefly**.

Slowtation — derogatory nickname for the Citation aircraft. Also **Nearjet**.

SLUF — Short Little Ugly Fucker: A7 attack bomber (U. S., Vought). —from Paul Turk.

2. The Douglas AC-7 Corsair 2 CV aircraft, the acronym standing for "Short Little Ugly Fellow," likely inspired by the **BUFF**.

Snake — nickname of the Bell AH1 **Cobra** helicopter.

Spad — "a small, World War I British biplane." —Berrey and Van den Bark, *The American Thesaurus of Slang*. The Spad, though widely flown by British and American pilots, was actually a French-built fighter and, due to its powerful His-pano-Suiza V-8 engine, swung the balance of power, for a time, in the Allies' favor. Also, in the U.S. Navy in Viet Nam, a name applied to the Douglas A-1 Skyraider, known to the USAF as **Sandy**.

Speedbird — British Airways (formerly British Overseas Airways Corporation) name for its planes.

spinning incinerator — irreverent nickname given the de Havilland DH-2 pusher biplane of WWI.

Spirit of Miami — A Boeing 727-100 sunk in the waters off Miami to form part of an artificial barrier. The aircraft, once owned by National Airlines and Pan American World Airways, flew commercially from 1965 to 1991. It was repainted in the liveries of Eastern Airlines and Pan Am, both now defunct Miami-based air carriers, and christened Spirit of Miami before being lowered to its watery grave.

Spirit of St. Louis — the Ryan NYP monoplane flown by Charles Lindbergh in his historic solo transatlantic flight to Paris in 1927.

Spirit of the Valley of the Moon — the grand appellation given the first DH-60 Moth in Canada. It was bought by the Department of Marine Fisheries as a reconnaissance plane for the Hudson Strait Expedition of 1927.

Spit — Supermarine Spitfire WWII British fighter.

Spooky — AC-47, Vietnam, also known as **Puff the Magic Dragon.**

Spruce Goose — giant wooden flying boat flown, for 70 seconds, once, by Howard Hughes, over Long Beach Harbour, California. With a 97-metre wingspan, 68 metres long, the Hughes-Kaiser HK-1 eight-engine flying boat was designed by Hughes in 1942 in partnership with Henry J. Kaiser. The HK-1 was intended as a transatlantic cargo and troop carrier. The hull was over 200 feet long and could carry a 60-ton tank or up to 700 troops. Plagued by design problems, Kaiser and the US government withdrew support in 1944 and Hughes, carrying on alone, finished building the prototype in 1947, taking the bird aloft for its one and only flight. Popularly thought to be the world's largest aircraft ever, it was such until Lockheed's C5 Galaxy.

Starbarge, Starpig — derogatory nicknames for the Beech Starship, notably unsuccessful aircraft of the early 1990s. "Beech's Edsel."

state bird of Alaska — the Cessna 172.

state bird of Texas — Southwest Airlines Boeing 737.

Stealth Bomber — Northrop B2 low-profile covert jet bomber.

Stealth Fighter — Lockheed F117 Night Hawk.

stickleback — the anti-submarine version of the Vickers Wellington Mark VIII.

St. Louis Slugger — the McDonnell F-4 jet fighter, also known as the "Double Ugly."

Stormovik — Ilyushin IL2 single-engine Russian ground attack aircraft of WWII. Known as "tank killer" or "tank buster." German ground forces called it **Black Death**; Russian pilots dubbed it **Flying Tank** and **Hunchback.**

Stringbag — Fairey Swordfish. Massive biplane, single-engine radial. WWII 110-mph torpedo bomber. Name came from the "wire bracing of her wings." — Hornick, *Famous Aircraft of the World*. The appellation derives from the aircraft's versatility which became apparent during test flights. "...but when trying it out with all those different loads some wag remarked, 'No housewife on a shopping spree could cram a wider variety of articles into her stringbag.' The name stuck and from that moment the pilots always called it the Stringbag when talking about it among themselves." — Lamb, *War in a Stringbag*.

Stuka — short for Sturzkampfflugzeug (German for "dive warplane"). Specifically the Ju-87 but generically "a term descriptive of all dive-bombers." — Green, *Warplanes of the Second World War*. Called "as much a product of propaganda as accomplishment" because early in its career it was employed against inferior forces. Also known as **Tank of the Sky**. Circa 1939.

Super Guppy — see **Airbus**.

Super Jolly MH-53 — "Helicopter used in rescue operations from enemy territory (UK source)" — John and Adele Algeo, "Among the New Words: Gulf War Glossary," *American Speech* (Winter 1991).

Super Scooper — "That's the name the California news media has fondly attached to the Canadair CL-215/215T/415," the water bomber. —*Canadair News*, March/April 1994. The aircraft can scoop 6,130 litres of fresh or salt water in 12 seconds, then "roar off on a ridge-hopping, canyon-diving attack on brush fires." See also **le Canadair, pelican**.

Swiss Submarine — Swiss-converted post WWII P 16 fighter-bomber. Two of the six prototypes ended up at the bottom of Lake Constance, yet the project contrib-

uted to jet-wing research, leading to the Learjet. — Szurovy

Swordfish —name for the Fairey torpedo bomber of WWII, known as the **Stringbag**.

Tadpole — nickname for the USN Grumman A-6 Intruder used in Viet Nam. See **Frying Pan**.

Tank of the Sky — JU88 **Stuka**.

Tante Ju — see **Auntie Ju**.

T-bird — "TP 80C, redesignated T-33A," first appeared 1947. —Stevens, *"There I was....."*

T-Craft — the Taylorcraft, popular 1930s forerunner of the Piper Cub, first built by Taylor Aircraft, in a partnership between Taylor brothers and William Piper.

Texan — North American T-6 WWII trainer. See also **Harvard, Yale**.

Thud — short for Thunderthud, officially Republic F-105 Thunderchief, fighter-bomber, Vietnam era. It's been said that it was nicknamed "for the sound which it made if it went too slow."

Tiffie — nickname for the Hawker Typhoon, RAF early WWII biplane fighter.

Tin Goose — nickname for the Ford Trimotor transport of the

1920s, "biggest available craft in Canada in 1935."

Tin Lizzie — "a Ford aeroplane." — Weseen, *A Dictionary of American Slang*.

Tin Parachute — nickname for the Aerospatiale Rallye sportsplane because of its characteristic 750 fpm nose-high descent rate at low speeds.

Tommy — nickname for the Thomas-Morse S-4 biplane trainer of 1917. The Tommy gained notoriety playing British, French and German aircraft in postwar flying films.

Tomorrowhawks — irreverent name given by troops in Papua, New Guinea, for **Tomahawks** very late in arriving. Also **Mythhawks**.

Tripehound — nickname for the Sopwith Triplane fighter of WWI; also applied to the Fokker Dr-1 triplane.

Troopship — military version of the Fokker F-27 Friendship, c. 1955. Designated F-27M.

Tunnan — "barrel", in Swedish, nickname for the Saab J-29 jet fighter because of its plump fuselage shape.

Tweetybird — nickname for the Cessna T-37 twin-jet side-by-side trainer of the late 1950s. Also known as the **Converter**.

Twin Bee — United Consultants' twin-engine conversion of the Republic **Sea Bee**. The UC1 Twin Bee utilized the basic Sea Bee airframe and hull. The notoriously underpowered Franklin pusher is replaced with two 100-HP Lycoming tractors. According to *Air Progress Magazine*, "entry to the cabin is via two large suicide doors, which opened into the airstream. Should one of these babies come open in flight, it would be blown back into the propeller, and I doubt the accident would be survivable, unless the pilot and occupants were very lucky."

Two Fighters with one Pilot — Japanese name for the Lockheed P-38 **Lightning**.

U-bird — nickname for the Chance Vought F-4U Corsair flown by the USMC during Korea. Also, see **Skyknight**.

Ural Bomber — German WWII design program for four-engine long-range strategic bombers.

Vee-strutter — nickname for the Albatros D-III and D-V because of the shape of its struts.

Vibrator — nickname for the Vultee BT-13 Valiant trainer of WWII. "This was mainly because of their characteristic shudder in a stall or a spin. The **pitot** head was mounted on the outboard left wing on the end of about a four-foot tube. In a stall or spin this tube would oscillate through about a two-foot arc." — Bill Kennedy Jr., *Air Classics Magazine*, May 1980. Also known as a **widowmaker**.

Victory Bomber — concept for a 1940 RAF bomber weighing 50 tons, capable of carrying a ten-ton bomb 4,000 miles at 320 mph at 45,000 feet, drawn up by Vickers designer Barnes Wallis, but never built.

Viggie — USN nickname for the North American Rockwell A-5 Vigilante Mach 2 carrier-borne bomber/recce designed to eject a nuclear weapon out the rear of the fuselage through a tunnel between the two engines, used in Viet Nam.

Vomit Comet — aircraft used for training in weightlessness — USAF C-135. —from Paul Turk.

Wagabond — Piper Vagabond high-wing single-engine fabric-covered lightplane of the 1940s. Notoriously underpowered. Known as the **Fat Cub**. Recently an improved kit-built version called the **Wagabond** has been offered by Wag-Aero.

Wal — "whale," Dornier eight-engine flying boat, Luftwaffe, WWII.

Warthog — nickname given the Fairchild A-10 ground support aircraft, some say because it is so ugly. It was so named especially for its snub nose, according to John and Adele Algeo, "Among the New Words: Gulf War Glossary," *American Speech* (Winter 1991).

Weasel — a fighter aircraft equipped to ferret out and destroy enemy radar and missile installations in Viet Nam. —Chuck Yeager's autobiography.

Whale — nickname for the Douglas EKA-3B Skywarrir, carrier-borne USN Viet Nam electronic intelligence, tanker, **recce**, and sometimes strike aircraft, because of its towering size. The USAF operated them as B-66 Destroyers.

Whispering Death — name given by the Japanese to the Bristol Beaufighter because its twin radial engines were so quiet they couldn't hear them coming.

Whistling Death — Japanese name for the Chance Vought F-4U Corsair during WWII. Sometimes also called by them, "Whispering Death."

Whisperjet — Eastern Air Lines name for its Boeing 727s, when they were first introduced. Recently, a nickname applied by one journalist to the Canadair Regional Jet, much more appropriately than to the Boeing, though from the point of view of image, the phrase is "used," out of date.

Widowmaker — may be used for any craft prone to crash or difficult to fly, having been applied to the Martin B-26 **Marauder**, Vultee BT-13 **Vibrator** and the Lockheed F-104 **Starfighter**. It was early attached to the Curtiss P-40, which General Chennault's **Flying Tigers**, in the Orient, did in fact make perform well, using ace pilots.

Wild Weasel — combat aircraft equipped with radar jamming and air-ground rockets to knock out surface-to-air missile sites, initially applied to USAF Republic F-105 Thunderchiefs, Gulf War.

2. Gulf War missile, from its tail-code designation WW.

Wimpey — from Popeye comic strip (his chubby pal was named J. Wellington Wimpey), the Vickers Wellington bomber of WWII, geodesic design, sometimes called "portly."

Wobbly Goblin — nickname of the Lockheed F-117 **Night Hawk**.

Wooden Wonder — RAF nickname for the de Havilland Mosquito of WWII, because of its almost entirely wooden construction. Affectionately known as **Mossie, Freeman's Folly**.

Work Horse — original name of the Piasecki (later Vertol) YH-21 Shawnee. When the helicopter went into production in 1955, Piasecki was no longer with the firm, which had been renamed Boeing-Vertol.

Yale — Canadian name for the North American T-6 **Texan**, known to the British as **Harvard**.

Yellow Peril — nickname for the Stearman N2S Kaydet, so dubbed by US naval aviation cadets because of its propensity to **ground loop** and because they were painted bright yellow.

Some printed sources

AAF: The Official Guide to the Army Air Forces. Pocket Books, 1944.

Air & Space Magazine,

Air Classics Magazine.

Airliners Magazine.

Algeo, John and Adele. "Among the New Words: Gulf War Glossary," *American Speech* (Winter, 1991), pp. 380-406.

———. "Among the New Words: Gulf War Supplement," *American Speech* (Spring, 1992), pp. 83-93.

Allward, Maurice. *An Illustrated History of Seaplanes and Flying Boats.* Dorset, 1981.

Barlay, Stephan. *The Final Call.* London, Sinclair-Stevenson, 1990.

Baxter, Gordon. *More Bax Seats.* 1988.

Beaudoin, Ted . *Walking on Air.* Vernon, BC: Paramount House, 1986.

Berger, Monty and Brian Jeffrey Street, *Invasions without Tears: The Story of Canada's Top-scoring Spitfire Wing in Europe during the Second World War.* New York: Random House, 1994.

Berrey, Lester V. and Melvin Van den Bark. *The American Thesaurus of Slang.* New York: Crowell, 1947.

Borins, Sanford F. *The Language of the Skies: The Bilingual Air Traffic Control Conflict in Canada.* Kingston: McGill-Queen's, 1983.

Bowen, Ezra. *Knights of the Air.* Alexandria, VA: Time-Life Books, 1981.

Brickhill, Paul. *The Dam Busters.* London: Pan, 1951.

Caiden, Martin, "Doolittle's Raid on Tokyo," in *Fighting Eagles,* ed. Phil Hirsch. New York: Pyramid Books, 1961.

Callander, Bruce D. "Jargon of the Air," *Air Force Magazine* (October, 1992), pp. 52-55.

Canadian Forces Aircom Regulation Glossary (1973).

Cannon, Hardy and Bill Stratton. *Box Seat Over Hell.* San Antonio, TX: Morton Printing, 1985.

Capt. X. *Safety Last.*

Clark, Alan. *Aces High.* Fontana/Collins, 1973.

Cleveland, Carl M. *Boeing Trivia.* Seattle: CMC Books, 1989.

Contact, newsletter of the Commonwealth Air Training Plan Museum (Brandon, Man.)

Coonts, Stephen. *Flight of the Intruder.* Annapolis: United States Naval Institute, 1986.

The Minotaur. New York: Dell, 1989.

Cushing, Steven. *Fatal Words: Communication Clashes and Aircraft Crashes.* Chicago: University of Chicago Press, 1994.

Deighton, Len. *Fighter.* St. Albans, Hertz.: Triad/Panther, 1979.

Goodbye, Mickey Mouse. London: Hutchison & Co., 1982.

Drew, Lt. Col. George A., *Canada's Fighting Airmen.* Toronto: MacLean, 1930.

Dugal, Xelphin V. *Readings on Aviation.* (1929).

Dunmore, Spencer. *Ace.* Totem, 1983.

Dunmore, Spencer and William Carter, *Reap the Whirlwind.* Toronto: McClelland and Stewart, 1991.

Ethell, Jeffrey. *Total Force.* Charlottesville, VA: Thomasson-Grant, 1988.

Forrester, Larry. *Fly For Your Life.* William Collins and Son, 1960.

Foster, J. A. *Sea Wings.* Methuen, 1986.

French, Edward B. *A Dictionary of Aeronautics, with Glossaries of Aerological and Navigational Terms.* Chicago: Mentzer, Bush & Co., 1945.

Gallico, Paul. *The Hurricane Story.* Berkeley Medallion, 1959.

Garrison, Paul. *Illustrated Encyclopedia of General Aviation.* Tab, 1990.

Green, William. *The Observer's World Aircraft Directory.* Frederick Warne & Co., 1961.

Gunston, Bill, ed. *Chronicle of Aviation.* London: Chronicle Communications Ltd., 1992.

Halberstadt, Hans. *Airborne: Assault from the Sky.* Novato, CA: Presidio Press, 1988.

Hamann, Fred. *Air Words.* Seattle: Superior, 1945.

Hamilton, Tim. *The Life and Times of Pilot Officer Prune: Being the Official Story of Tee Emm.* London: His Majesty's Stationery Office, 1991.

Harding, Stephan. *U S Army Aircraft since 1947.* Stillwater MN: Specialty Press, 1990.

Heath, Layne. *CW2*. New York: William Morrow, 1990.

Heflin, Woodrow Agee, ed. *The United States Air Force Dictionary*. Air University Press, 1956.

Helm, Eric. *The Raid*. Worldwide Library, 1988.

Hirsch, Phil, ed. *Fighting Eagles*. New York: Pyramid Books, 1961.

Hooper, Bill. *Pilot Officer Prune's Picture Parade*. London: HMSO Publications, 1991.

Hotson, Fred W. *The DH Canada Story*. de Havilland Aircraft of Canada, Ltd., 1978.

HW. *What's the Gen?* London: John Crowther, Ltd., 1942. A copy annotated by hand by Jean Elizabeth Wall Kinloch, WVS, just after publication, is in the archives of the Yorkshire Air Museum at Elvington, near York. Another is at the British Commonwealth Air Training Plan Museum in Brandon, MN.

ICAO 50th anniversary book.

Infield, Glen, "The Day Ted Williams Almost Got It," in *Fighting Eagles*, ed. Phil Hirsch. New York: Pyramid Books, 1961.

Johnson, Robert S. *Thunderbolt*. Ballantine, 1958.

Jones, Paul. *An Alphabet of Aviation*. Philadelphia: Macrae Smith Co., 1928.

Jordanoff, Assen. *Jordanoff's Illustrated Aviation Dictionary*. New York: P. F. Collier & Sons, 1942.

Keith, Ronald A. *Bush Pilot with a Briefcase: the happy-go-lucky story of Grant McConachie*. Toronto: Paperjacks, 1972.

Lamb, Charles. *War in a Stringbag*. Arrow Books, 1977.

Langdon, David. *"All Buttoned Up!" A Scrapbook of R.A.F. Cartoons*. London: The Sylvan Press, n. d.

Logie, Stuart. *Winging It: The Story of the Canadair Challenger*. Toronto: Gage, 1992.

Lopez, Barry. "On the wings of commerce," *Harper's* (October, 1995).

McIntosh, Dave. *Terror in the Starboard Seat*. Markham, ONT: Paperjacks, 1981.

McVicar, Don. *The Grass Runway*. Dorval, Quebec: Ad Astra, 1991.

Mersky, Peter B., and Norman Polmar. *The Naval Air War in Viet Nam*. New York: Zebra Books, 1981.

Mitchell, Jared. "Swan Song for a Workhorse," *Report on Business Magazine* (April 1996).

Molson, K. M. *Pioneering in Canadian Air Transport.* (1974).

Munro, Raymond Z. *The Sky's No Limit.* Toronto: Key Porter, 1985.

Murray, Thomas E. "The Language of Navy Fighter Pilots", *American Speech* 61:2 (Summer 1986), pp. 121-129.

News Aloft: American Airlines Flight Service Procedures (in-house newsletter).

Ocean, E. B. *Dictionary of Air Transport and Traffic Control.* London: Granada, 1984.

Pagé, Victor W. and Paul Montariol. *Glossary of Aviation Terms/ Termes d'Aviation.* New York: Norman W. Henley, 1917.

Parkin, Tom. *WetCoast Words.* Victoria, BC: Orca, 1989.

Partridge, Eric. *A Dictionary of R. A. F. Slang.* London: Michael Joseph, 1945.

Pierce, Robert Morris. *Dictionary of Aviation.* New York: Languages Publishing Co., 1914.

Porter, Col. R. Bruce. *Ace! A Marine Night-Fighter Pilot in World War II.*

Poteet, Lewis and Jim Poteet. *Car & Motorcycle Slang.* Ayers Cliff, Que.: Pigwhistle, 1992; revised edition *Car Talk* published by Montreal: Robert Davies Inc., 1997.

Private Pilot Magazine.

Raff and Anthony Armstrong, *Plonk's Party.* London: Methuen, 1942.

Rich, Ben R. and Leo Janos. *Skunk Works: A Personal Memoir of My Years at Lockheed.* Boston: Little Brown, 1994.

Rose, Howard N. *A Thesaurus of Slang.* New York: Macmillan, 1934.

Sabbagh, Karl. *Twenty-First-Century Jet: the making and marketing of the Boeing 777.* New York: Scribner, 1996.

Searls, Hank. *The Crowded Sky.* Harper and Row, 1960.

Spring, Joyce. *Daring Lady Flyers.* Porters Lake, NS: Pottersfield Press, 1994.

Stevens, Bob. "*There I was...flat on my back*". Blue Ridge Summit,PA: Aero, Tab Books, 1975.

Stevens, Bob. *If You Read Me, Rock the Tower.* Blue Ridge Summit, PA: Aero, Tab Books, 1980.

Stoffey, Col. Bob. *Cleared Hot! A Marine Combat Pilot's Vietnam Diary.* New York: St. Martin's Press, 1992.

Swanborough, F. G. *Military Transports & Training Aircraft of the World*. Temple Press Books, 1965.

Szurovy, Geza. *Learjets*. Osceola, WI: Motorbooks, 1996.

Thurston, Arthur. *Bluenose Spitfires*. Hantsport NS: Lancelot, 1979.

U.S. Aviator Magazine.

U. S. News & World Report, *Triumph without Victory*. 1992.

Ventou-Duclaux, L. *L'Aviation expliquée*. Paris: F.-Louis Vivien, 1909.

Ventou-Duclaux, L. *Petite Encyclopédie Aéronautique*. Paris: F. Louis Vivien, 1910.

Wallis, F/O Sydney. "Logbook," unpublished ms. made available by Mrs. Alixe Wallis, Whiterock BC, Canada, through CBC television series "No Price Too High."

Ward-Jackson, Squadron Leader C. H. *It's a Piece of Cake or R.A.F. Slang Made Easy*. London: The Sylvan Press, n.d.

Yeager, General Chuck and Leo Janos. *Yeager*. Toronto: Bantam, 1985.

Weseen, Maurice H. *A Dictionary of American Slang*. New York: Crowell, 1934.

Williams, Henry Lionel. *Casey Jones Cyclopedia of Aviation Terms*. New York: McGraw-Hill, 1946.

Williams, James N. *The Plan: Memories of the British Commonwealth Air Training Plan*. Canada's Wings, 1984.

Wilson, George C. *Super Carrier*. New York: Berkley Books, 1986.

Wings (TV documentary series) Discovery Channel.

Wragg, David W. *A Dictionary of Aviation*. New York: Frederick Fell, 1974.

Zweng, Charles A. *Encyclopedia Aviation Dictionary*. Los Angeles: Pan American Navigation Service, 1974.

Thanks to Bob Parke (McGraw-Hill Show Dailies editor), Paul Turk (*Avmark*), Catherine Chase (Canadair), Richard Gordon (Shorts), Grant McLaren (Aviation writer for *Professional Pilot* and others), several anonymous informants in the airline and aircraft industries, and some who may be named, like John Cavill, formerly PR for Air Canada's eastern Canadian region, Jennifer Sherwen (Swissair), Doug Sample CD of the Canadian Branch, Yorkshire Air Museum, Elsa Schieder, Peter Otto and Bob Wohl of Canadair, Murray and Jean Kinloch of University of New Brunswick, Ron Cosper of St. Mary's University (Halifax), Pat Dudley of Special Forces, Jeff Miller, Dave Monasmith, Bill Robinson, and Dick Etherington of Learjet, Bill Stratton, Ed Hirsch of International Aviation Magazine Group, Eric Hehs of *Code One* magazine of Lockheed Fort Worth and John Wheeler of Boeing; Maureen Charnaux and James Occhipinti ("the token pilot on the Hill") of Peter Pappas' staff at AMR (American Airlines, Fort Worth); Richard Fields, MD11 flight instructor with American Airlines; Larry Sall (University of Texas at Dallas special collections library director); Don Hackett of Dawson College, Montréal; Sean Kelly, USAF KC-135 tanker pilot; Ray Seale, former 'Nam Huey pilot, now on the Internet; Susan Kazenel for clippings and tips; John David, aviation mechanic and airshow fan.